Computer Simulation of Carcinogenic Processes

Editor

B. D. Silverman, Ph.D.
Research Staff Member
IBM Thomas J. Watson Research Center
Yorktown Heights, New York

CRC Press, Inc.
Boca Raton, Florida

Library of Congress Cataloging-in-Publication Data
Computer simulation of carcinogenic processes.

Bibliography: p.
Includes index.
1. Carcinogenesis—Computer simulation.
I. Silverman, B. D.
RC268.5.C645 1987 616.99'4071 87-23846
ISBN 0-8493-6539-2

Direct all inquiries to CRC Press, Inc., 2000 Corporate Blvd., N.W., Boca Raton, Florida, 33431.

International Standard Book Number 0-8493-6539-2

Library of Congress Card Number 87-23846
Printed in the United States

PREFACE

The computer, whether used to simulate complex biochemical processes, graphically illustrate certain essential factors, or keep track of relevant genetic information, has become an important tool in cancer research. The present volume is a collection of six articles which describe how the computer has enabled one to check out hypotheses concerning the induction or initiation of cancer. Three of the articles are related to the examination of hypotheses concerning certain details of the protein structures inferred from oncogenic sequences. The remaining three articles examine the properties of an extensively investigated group of carcinogens, the polycyclic aromatic hydrocarbons, in an attempt to infer how their intrinsic chemical properties as well as their ability to interact with DNA, contribute to the initiation step of carcinogenesis.

THE EDITOR

B. D. Silverman, Ph.D., is a Research Staff Member at the IBM Thomas J. Watson Research Center, Yorktown Heights, New York. Dr. Silverman received his Ph.D. in theoretical solid state physics at Rutgers University in 1959. He was a member of the theory group at the Raytheon Research Division, Waltham, Massachusetts until 1966. He then joined The NASA Electronics Research Center, Cambridge, Massachusetts, where he became manager of the Advanced Research Department. In 1969 he joined the IBM Research Laboratory in San Jose, California, where he managed a group responsible for developing materials for optical computer memories. He transferred to Yorktown Heights in 1973 managing a group involved with the physics of conducting organic materials.

Dr. Silverman has worked in such divergent areas as ferroelectrics, ultrasonics, quantum electronics, optical computer memories, organic metals, chemical carcinogenesis, and polymers. During 1982 he spent a sabbatical year as manager of IBM's Toxicology Group in San Jose, California. He is a fellow of the American Physical Society.

CONTRIBUTORS

Winona C. Barker, Ph.D.
Associate Director of Research
Protein Identification Resource
National Biomedical Research Foundation
Washington, D.C.

Suse Broyde, Ph.D.
Associate Professor
Department of Biology
New York University
New York, New York

Jean Garnier, D.Sc.
Director of Research
Laboratory of Biochemistry-Biophysics
University of Paris-Sud
Orsay, France

David G. George, Ph.D.
Senior Research Scientist
Protein Identification Resource
National Biomedical Research Foundation
Washington, D.C.

Brian E. Hingerty, Ph.D.
Member, Research Staff
Health & Safety Research Division
Oak Ridge National Laboratory
Oak Ridge, Tennessee

Lois T. Hunt, Ph.D.
Senior Research Scientist
Protein Identification Resource
National Biomedical Research Foundation
Washington, D.C.

Reiko Kuroda, D.Sc.
Associate Professor
Department of Chemistry
College of Arts and Sciences
The University of Tokyo
Komaba, Meguro
Tokyo, Japan

Stephen Neidle, Ph.D.
Director
CRC Biomolecular Structure Unit
Institute of Cancer Research
Sutton, Surrey, England

Matthew R. Pincus, M.D., Ph.D.
Department of Pathology
New York University Medical Center
New York, New York

Barry Robson, D.Sc.
Theoretical Biochemistry Laboratory
University of Manchester Medical School
Manchester, England

B. D. Silverman, Ph.D.
Research Staff Member
IBM T. J. Watson Research Center
Yorktown Heights, New York

TABLE OF CONTENTS

Chapter 1

HOMOLOGY AND SIMILARITY IN AMINO ACID SEQUENCES: DETECTION, ANALYSIS, AND INTERPRETATION

Winona C. Barker, Lois T. Hunt, and David G. George

TABLE OF CONTENTS

I. INTRODUCTION

The recognition of structural, functional, or evolutionary relationships among proteins often provides fundamentally new insights into cellular mechanisms and can also lead to a more detailed understanding of the breakdown of these mechanisms in disease states such as cancer. The recent discoveries that viral oncogenic proteins are related to growth factors,[1,2] to growth factor receptors,[3-5] and to the tyrosine-specific protein kinases[6-8] have together contributed to an emerging picture of the general mechanism for the stimulation of cellular proliferation. Disruption of this normal mechanism at any one of a number of points can induce transformation to the neoplastic phenotype.[9] Cellular replication is stimulated by exogenous growth factors, which according to the autocrine hypothesis may be secreted by the transformed cells themselves. The growth factors bind to specific receptors that are transmembrane proteins whose cytoplasmic domains usually have tyrosine kinase activity. Binding of growth factor induces the kinase activity, which triggers the intracellular events that eventually lead to DNA synthesis and mitosis. The inner (cytoplasmic) surface of the cell membrane may be the site of action of a number of oncogenic proteins,[10] including those that have structural homology with the tyrosine kinases (e.g., *src, yes, abl, fgr, etc.*). The *ras* proteins function at a simlar location. These bind and hydrolyze GTP and may be related to proteins that relay signals from certain hormones to the interior of the cell. Certain other oncogene products (*myb, myc, fos*) act in the nucleus and are known to bind to DNA. The fact that all identified proto-oncogenes are well conserved in evolution supports the hypothesis that they play critical roles in normal growth processes.

Frequently, the first step in the discovery of such interesting protein relationships is a search through a database of known protein sequences. The similarities detected may be ''local'', involving limited regions of the sequence, or ''global'', involving the entire sequence. The common searching algorithms are designed to find regions of local similarity; other methods must be employed to determine the extent of the regions of similarity. Proteins may show homology only in limited regions or domains rather than along their entire lengths. This has made the problem more interesting although it necessitates a more careful look at the data.

II. SEARCHING FOR SEQUENCE SIMILARITIES: GENERAL CONSIDERATIONS

A. Identities in Protein Sequences

Although the general purpose of searching the protein sequence database is to find proteins that are similar to a test sequence, the appropriate searching techniques and subsequent evaluation of the results are dependent on the needs and interests of the researcher. All of the methods to be described here produce alignments of the test sequence (or a portion of it) with selected sequences (or subsequences) from the database.[11,12] Amino acid sequences are aligned by pairing their consecutive amino acids. An "identity" is the occurrence of the same amino acid at corresponding positions in both sequences. The simplest kind of search is to look for strings of uninterrupted identities. Long strings of identical amino acids occur in closely related proteins (those that are greater than 80% identical). These may be encoded by recently duplicated or allelic genes within an organism, or by genes for the same protein in related organisms, or even by highly conserved genes in very distantly related organisms. Because there are mechanisms for maintaining the close similarity of duplicated genes within a genome[13-15] and selection may maintain a protein sequence almost unchanged throughout a long span of evolution,[16] proteins may be nearly identical even though their genes diverged some time ago.

Closely related proteins generally exhibit very similar secondary and tertiary structures.[17] (Secondary structure refers to the helices, turns and strands occurring in the molecule; tertiary structure refers to the characteristic, though not rigid, three-dimensional shape formed by the interactions of these elements.) Antigens that react with a protein may cross react with the identical or nearly identical regions of closely related proteins. The presence of a short stretch of four to six identical amino acids in unrelated proteins, however, is not necessarily indicative of antigenic cross reactivity or similarity of secondary structure. Identical spans of such length can be found forming different types of secondary structures in different proteins.[18]

In distantly related sequences (about 50 to 80% different), there may be regions containing spans of uninterrupted identities. Such regions may have been well conserved within several groups of proteins because they play some essential role in the functional activity or in the structural integrity of the proteins. Except in such regions, the identities in distantly related proteins may be sufficiently scattered so that searching for strings of uninterrupted identities is not a productive strategy. Instead one may look for longer regions in which the number of identical amino acids is much greater than expected. It is unusual to find unrelated proteins with regions of 40 to 50% identity over spans of 25 or more amino acids unless the regions are quite restricted in the types of amino acids that occur within them.

B. Gaps in Aligned Sequences

In comparing distantly related sequences, the number of identities can often be substantially increased by inserting "gaps" in the sequences. These gaps may represent evolutionary processes such as the insertion or deletion of genetic material; occasionally they represent sequencing errors. Unfortunately, the number of identities seen between unrelated sequences is also greatly increased by allowing gaps in the alignment.[19] For example, in aligning randomly generated 20-residue sequences, the probability of finding five identities was increased tenfold by allowing one gap to be introduced; in alignments of 100-residue sequences, the mean number of identities approximately doubled when one to five gaps of only one amino acid each were inserted.[20] Great care must be taken not to overinterpret the significance of identities observed after gaps have been inserted. Most searching methods, however, do not allow gaps because that would increase to an unacceptable level the time needed to do the search. Researchers who are not constrained by such limitations may wish to explore the methods of Goad and Kanehisa.[21,23]

C. Similarity Scoring Matrices

When related sequences are aligned, it is often observed that nonidentical amino acids at corresponding positions are chemically similar. On the assumption that similar amino acids

	C Cys	S Ser	T Thr	P Pro	A Ala	G Gly	N Asn	D Asp	E Glu	Q Gln	H His	R Arg	K Lys	M Met	I Ile	L Leu	V Val	F Phe	Y Tyr	W Trp
C Cys	12																			
S Ser	0	2																		
T Thr	-2	1	3																	
P Pro	-3	1	0	6																
A Ala	-2	1	1	1	2															
G Gly	-3	1	0	-1	1	5														
N Asn	-4	1	0	-1	0	0	2													
D Asp	-5	0	0	-1	0	1	2	4												
E Glu	-5	0	0	-1	0	0	1	3	4											
Q Gln	-5	-1	-1	0	0	-1	1	2	2	4										
H His	-3	-1	-1	0	-1	-2	2	1	1	3	6									
R Arg	-4	0	-1	0	-2	-3	0	-1	-1	1	2	6								
K Lys	-5	0	0	-1	-1	-2	1	0	0	1	0	3	5							
M Met	-5	-2	-1	-2	-1	-3	-2	-3	-2	-1	-2	0	0	6						
I Ile	-2	-1	0	-2	-1	-3	-2	-2	-2	-2	-2	-2	-2	2	5					
L Leu	-6	-3	-2	-3	-2	-4	-3	-4	-3	-2	-2	-3	-3	4	2	6				
V Val	-2	-1	0	-1	0	-1	-2	-2	-2	-2	-2	-2	-2	2	4	2	4			
F Phe	-4	-3	-3	-5	-4	-5	-4	-6	-5	-5	-2	-4	-5	0	1	2	-1	9		
Y Tyr	0	-3	-3	-5	-3	-5	-2	-4	-4	-4	0	-4	-4	-2	-1	-1	-2	7	10	
W Trp	-8	-2	-5	-6	-6	-7	-4	-7	-7	-5	-3	2	-3	-4	-5	-2	-6	0	0	17

FIGURE 1. The mutation data scoring matrix. (From Dayhoff, M. O., *Atlas of Protein Sequence and Structure*, Vol. 5, Suppl. 3, National Biomedical Research Foundation, Washington, D.C., 1979, 352. With permission.)

often can replace each other in a protein sequence with very little effect on the function or structure of the protein, two sequences can be quite similar even though they do not share a high percentage of identities. Therefore, some searching methods (and many of the other sequence comparison methods to be discussed herein) use a scoring matrix that assigns a value to each possible pair of amino acids. In a similarity scoring matrix, higher values are assigned to more similar pairs and lower values to dissimilar pairs. The simplest similarity matrix, which we call the unitary matrix, assigns values of +1 for identities and 0 for nonidentities. A scoring matrix based on the genetic code reflects the maximum numbers of nucleotides that the codons for two amino acids may have in common, for example, scoring +3 for identities, +2 for amino acids whose codons must differ by at least one nucleotide, +1 for those whose codons must differ by two nucleotides, and 0 for those whose codons cannot have any nucleotides in common. McLachlan[24,25] derived a scoring matrix based on the alternative amino acids found at corresponding positions in alignments of groups of related sequences. We have derived a scoring matrix (the mutation data matrix) based on amino acid replacements between closely related present-day sequences and those inferred as common ancestors on evolutionary trees.[26] The amino acids that did not change and the relative exposure of the sequences to mutational change were taken into account.

Although all of these scoring systems are effective for finding closely related sequences, the mutation data matrix is often best for detecting distantly related sequences.[27,28] This matrix (shown in Figure 1) clearly reflects the physicochemical properties of the amino acids. Figure 1 has been arranged to designate groups of chemically similar amino acids: the aromatic group (tryptophan, tyrosine, phenylalanine); the hydrophobic groups (valine, leucine, isoleucine, methionine); the basic group (lysine, arginine, histidine); the acid, acid-amide group (glutamine, glutamic acid, aspartic acid, asparagine); a group of amino acids that are not strongly hydrophilic or hydrophobic (glycine, alanine, proline, threonine, serine); and cysteine. Some groups overlap: the basic and the acid, acid-amide groups tend to replace one another to some extent, and phenylalanine interchanges with the hydrophobic amino acids more often than chance expectation would predict. These patterns are imposed principally by natural selection and only secondarily by constraints of the genetic code; they reflect the similarity of the functions of the amino acids in their interactions with one another in the three-dimensional conformation of proteins. Some of the properties of an amino acid residue that determine these interactions are size, shape, and local concentrations of electric charge; the conformation of its van der Waals surface; and its ability to form salt bonds, hydrophobic bonds, and hydrogen bonds.[26] For specific applications, it is possible to derive scoring matrices that specifically reflect one or another of these particular properties.

Many protein sequences are being inferred by translation of potential protein coding regions in nucleic acid sequences. Often the protein has not been isolated or its function is unknown. Possible activities of several types of viral oncogene products have been proposed after discovery of their homology with cellular proteins.[1-7] However, in many cases no relationship is found or a suspected relationship may be so distant that it cannot be demonstrated. Nevertheless, certain identifiable groups of proteins have similar properties that relate to functional similarities. For example, mitochondrial proteins and membrane-associated proteins tend to be hydrophobic, and many protease inhibitors are small molecules with a high content of functionally essential cysteine, a property that they share with certain growth factors and toxins. Searching the database for proteins that are structurally similar may provide testable hypotheses about the possible functional roles of a protein, leading to more rapid establishment of the specific roles of oncogenic proteins in carcinogenesis.

III. SEQUENCE-SEARCHING ALGORITHMS OF THE PROTEIN IDENTIFICATION RESOURCE

A. The Scan Command of PSQ

The protein identification resource provides several methods of searching large sequence databases.[29] The fastest way to identify identical or closely related sequences is the interactive Scan command of the protein sequence query (PSQ) program. This command searches the database for identically matching segments; because the command makes use of an indexed tripeptide table, the response is very rapid.[12] In the present database, a string of six consecutive amino acids is usually sufficient to identify a closely related protein (occasionally a string of five or six identities will be found in an unrelated protein also). Of the 64 million possible hexapeptides that can be formed from the 20 types of amino acids, only 512,700 actually occurred in the May 1985 database, which contained 694,014 amino acid residues in 3182 sequence entries. Table 1 shows the percentages of the possible peptides found for lengths up to ten.

As an illustration, the 20-nucleotide cDNA fragment AATTCTTCCTGCTGCTTTCC can be translated in each reading frame to yield the amino acid sequences NSSCCF, ILPAAF, and FFLLLS. With the Scan command, the user could rapidly ascertain that only the last hexapeptide is found in the protein database. One can also restrict Scan to search only for proteins containing the segment at the amino end by preceding the segment with the plus sign (+) or to search only for proteins containing the segment at the carboxyl end by following the sequence with an asterisk.

B. The Match Command of PSQ

For some applications it is desirable to be able to allow some mismatches between the test segment and segments retrieved from the database. The Match command of PSQ performs

Table 1
NBRF PROTEIN SEQUENCE DATABASE, RELEASE 5.0, MAY 15, 1985 (3182 SEQUENCES, 694,014 RESIDUES)

Length of peptide	Total possible N-peptides	Number found	Percentage found
N	20**N		
2	400	400	100.0
3	8,000	8,000	100.0
4	160,000	125,680	78.6
5	3,200,000	426,569	13.3
6	64,000,000	512,700	0.80
7	1,280,000,000	528,900	0.041
8	25,600,000,000	537,143	0.0021
9	512,000,000,000	543,302	0.00011
10	10,240,000,000,000	548,213	0.000005

```
EQHUA
Proenkephalin precursor - Human
Exact matches at         !!!!
Matching------------KRYGGFMKR
residue                 !                No.
     98  SKPEESHLLA KRYGGFMKR YGGFMKKMDE  0
    105  LLAKRYGGFM KRYGGFMKK MDELYPMEPE  1
    134  EEANGSEILA KRYGGFMKK DAEEDDSLAN  1
    184  DGSDNEEEVS KRYGGFMRG LKRSPQLEDE  2
    208  QLEDEAKELQ KRYGGFMRR VGRPEWWMDY  1
    228  GRPEWWMDYQ KRYGGFLKR FAEALPSDEE  1
    259  SYSKEVPEME KRYGGFMRF             2
```

FIGURE 2. Example of Match command. The human proenkephalin precursor was searched for the string KR(YGGF)MKR with no mismatches allowed for the subsequence enclosed in parentheses. There are seven occurrences of this string, corresponding to the enkephalin peptides and their flanking basic residues.

a direct search for segments that match a string of up to 30 amino acids with a specified number of mismatches.[30] The display (see Figure 2) shows the sequence position of the first residue in the matching segment, the ten residues preceding the matching segment, the matching segment, the ten residues following it, and the number of mismatches. The user may specify that certain residues must exactly match by enclosing them in parentheses. For example, when searching for KR(YGGF)MKR, mismatches are not allowed for residues YGGF. It takes several minutes to search the entire database using the Match command. The time can be reduced substantially by restricting the search to a subset of the database; e.g., one can search only the viral protein sequences.

The Match command can also be used to search a particular sequence for specific patterns. For example, one can locate the asparagine residues that are potential carbohydrate binding sites by looking for NXT and NXS in the sequence (X is an ambiguous amino acid symbol that designates any amino acid). As shown in Figure 2, the occurrences of a repeated pattern within a sequence can also be conveniently displayed (as long as no gaps are needed to align the repeat segments). As with the Scan command, + and * can be used to restrict the search to the amino or carboxyl ends of sequences. For instance, the search string +MXXXXXXXXX will cause the first 20 residues (10 in the matching segment and 10 following it) of all sequences beginning with methionine to be displayed.

C. The SEARCH Program

Program SEARCH compares a segment of a test sequence, typically 25 residues, with all segments of the same length in the database and with the progressively shorter segments at the ends of those sequences.[31] A similarity score is computed for each segment comparison by summing the amino acid pair scores for each of the aligned residues. The amino acid pair scores are defined by the similarity matrix. Up to 500 of the highest scoring segments are saved and a histogram of the distribution of all segment scores is displayed. Typically, for a 25-residue piece, the corresponding segments from all closely related sequences (up to 50% identical overall), and many from the more distantly related sequences, will appear above the distribution of segments from unrelated proteins.

The SEARCH program is well suited for use when only a partial or fragmentary sequence has been determined. Very commonly the amino ends of an intact protein chain and of its cyanogen bromide fragments will be subjected to sequenator analysis, producing sequences of segments 20 to 40 residues long. A SEARCH of the protein sequence database will easily identify the sequence if it is in the database. Moreover, a sequence that is within the same family (less than 50% different) would almost always be noticed in a search of five segments.[31] Inasmuch as sequences would have to change very rapidly in order for homologous proteins in different mammals to be more than 50% different, the presence of a homolog from any other mammal in the database would usually ensure the identification of the human sequence.[29] In the May 1985 database, there are 210 superfamilies of mammalian proteins. For some major taxonomic groups, however, the number of families and superfamilies represented is still quite small. The possibility of finding a related protein in the database depends on many factors, including the probability of success in searching for a single segment, the number of segments actually searched, and the number of representative types of proteins currently in the database. We have discussed these factors in detail elsewhere.[31]

Occasionally, the initial attempt to subject a newly isolated protein to sequenator analysis will yield two amino acids at each step because the sample is a mixture of two proteins or of two chains, each with a free amino end. A component of such a mixture can be easily identified if it is already in the protein sequence database. For instance, a mixture of prealbumin and alpha$_1$-antitrypsin would give the following information for each position:

1	2	3	4	5	6	7	8	9	10	11	12
Gly	Asp	Thr	Gly	Gly	Asp	Ala	Ala	Lys	Cys	Pro	Asp
Glu	Pro	Pro	Gln	Thr	Gly	Glu	Ser	Gln	Lys	Thr	Leu

A slight modification of the SEARCH program allows the database to be searched for all sequences that contain Gly or Glu followed by Asp or Pro followed by Thr or Pro, etc. If 12 positions are used, both prealbumin and alpha$_1$-antitrypsin would match perfectly and no other sequence currently in the database would match at more than eight positions.

In most cases, identification of only seven to eight amino acid residues (not necessarily contiguous) will uniquely identify a sequence if it is in the database. Thus, microsequencing of peptides containing certain radiolabeled amino acids can yield sufficient information to identify a protein. Such an identification made in the early stages of a sequence determination would save time and money by preventing redundant efforts.

D. The FASTP Program

FASTP is a new program (developed by Pearson and Lipman) for searching large protein databases.[32] It is an adaptation of the k-tuple search algorithm, described by Wilbur and Lipman.[33,34] that greatly improves its speed and sensitivity. The first stage of the algorithm finds the five best subsequence alignments of two sequences, allowing mismatches but no gaps. This part of the algorithm is very rapid but still considers all possible alignment registers and all possible lengths at a particular alignment register. For maximum speed, the algorithm considers only those identities that occur adjacent to another identity (k = 2), whereas for maximum sensitivity, all identities are considered (k = 1).

In the next stage of the algorithm, the five best subsequence alignments are scored using the mutation data matrix. For each sequence, the highest scoring alignment, called the ranking alignment, is selected. The corresponding subsequence score is called the ranking score. A histogram based on the ranking scores, and its mean and standard deviation, are displayed.

A number of highest ranking alignments are saved and realigned allowing gaps. An optimized score is computed from the new alignment. The optimization algorithm is equivalent to that of Needleman and Wunsch,[35] with the restriction that no contiguous string of gaps may be longer than a preset value, in this case 16. For related sequences, the optimized score is often much higher than the ranking score and the optimized alignment may involve a substantial fraction of the two sequences.

The method of choosing and scoring subsequence alignments is such that both very short stretches of high similarity (e.g., seven identities in a length of 10) and long stretches of lower similarity (e.g., 20 to 25% identity in a length of over 100) may occur among the top-ranking sequences. Only the latter are likely candidates for demonstrating evolutionary relationship; nevertheless, restricted regions of local similarity can be interesting from other points of view. It should also be stressed that the FASTP program does not display all regions of similarity between two sequences, but selects only the best. If, for example, two sequences are similar over their entire lengths except for a region in the middle that is very dissimilar or requires a large gap in one sequence to produce the proper alignment, only the amino-terminal or the carboxyl-terminal matching regions will be displayed, but not both. Similarly, the test sequence could be homologous with two regions of a sequence that has an internal duplication, but only one matching region will be found. Therefore, computer programs (such as those described below) must be used to give a more complete analysis of the similarity between the query sequence and the most interesting candidates found by these database searching methods.

IV. EVALUATION OF SEQUENCE SIMILARITIES

A. Introduction

Researchers commonly assert the biological significance of sequence similarities on the basis of one or more of the following: (1) near identity of the sequences; (2) statistical significance of the similarity based on various tests; (3) direct biological evidence. If an alignment of two sequences reveals near identity over a considerable length, the conclusion that the sequences are closely related seems clearly warranted. The biological problem is then to determine the origin and nature of the relationship: is it in fact the same protein, i.e., coded by the same gene; are they products of two very closely related genes that may be under separate control in different tissues; are they members of a multigene family in which gene conversion has kept the sequences very similar; has a gene been translocated from the host to a viral genome or vice versa, or from the nuclear genome to that of a cytoplasmic organelle? On the other hand, the assertion that proteins have a common origin or functional relationship solely on the basis of several short regions of similar sequence is unwarranted.

The statistical tests generally employed involve computing some measure of similarity for a large number of pairs of sequences; from this distribution a mean and standard deviation are calculated. A "Z value", expressed in SD units, can be calculated by dividing the difference between the observed value and the mean by the standard deviation. If the population of similarity scores is assumed to conform to a normal (Gaussian) distribution, then these Z values can be converted into probabilities using standard tables. Unfortunately, the assumption that scores from unrelated sequences are normally distributed is usually unjustified. Thus, standard statistical methods cannot be used to assess the significance of the results of the search.

B. Distribution of Scores from Database Searches

We have studied the distributions of scores from the SEARCH and FASTP programs and find that both have an excess of high scores. Probabilities based on Z values calculated from such distributions would overestimate the significance of the similarity. Even if distributions of scores from searching a large database were shown to be normal, the total number of comparisons made must be taken into account when the results are evaluated. If the number of independent comparisons made in doing the search is 1 million, then at least one score corresponding to a Z value of 4.75 SD is to be expected. Because the SEARCH program compares a fixed length of sequence, one can count the number of such comparisons made in doing the database search. However, these comparisons are not all independent because some of the comparisons overlap and because some of the database sequences are closely related to one another. The FASTP algorithm detects similarities spanning any length, making it difficult to estimate the effective number of comparisons made.

The database contains sequences with regions of very unusual composition. These may score very highly against a test sequence with a similar unusual distribution of amino acids and also give an overestimation of significance. For example, when we used FASTP to search a Balbiani ring protein against the database, alignments with more than 25% identities over lengths of more than 100 residues were found with rabbit and sea urchin histone H1, which also contain an unusually high percentage of lysine. When, as a control, the Balbiani ring sequence was shuffled and the search repeated, six of the ten best alignments were again with histone H1 and related sequences. The distribution of scores was very skewed, with 11 sequences attaining Z values of 6 SD or more. Another factor that may contribute to the inappropriateness of the normal distribution as a statistical model for database searches is the heterogeneity of the sequences in the database with respect to length and amino acid composition.

It is our experience that there is no single numerical criterion by which one can evaluate the biological significance of sequence similarity. We view with skepticism the attempts to assess probabilities based on database searches. Sequence comparison methods are best used as tools to detect possible relationships between sequences rather than to prove that they are related. The biological significance of sequence similarities can only be determined experimentally. With frequent use of a particular searching algorithm, including the searching of shuffled sequences as controls, one develops a sense of when a similarity may be unusual. As one of several guides to evaluating a FASTP search, Lipman and Pearson[32] suggest that Z values of ten or more are significant, those between six and ten are probably significant, and those between three and six are only possibly significant.

C. Other Methods for Analyzing Sequence Similarity

1. ALIGN and RELATE

Program ALIGN, which is based on the algorithm of Needleman and Wunsch,[35] determines a highest scoring alignment (including gaps) of two sequences (see Figure 3). The score is then compared with the highest possible scores obtained by aligning pairs of permuted sequences having the same amino acid composition and length as the two real sequences. The scores of the permuted sequences form a distribution for which the mean and standard deviation are calculated. The alignment score is the number of standard deviations (SD units) by which the score for the best alignment of the real sequences exceeds the mean score for the best alignments of the permuted sequences. Although we and others have in

```
Mutation Data Matrix (250 PAMs)
Bias= 6, Penalty = 6
WMVZ9      140 residues
19K protein - Vaccinia virus
WFHU1      160 residues
Transforming growth factor type 1 precursor - Human

          1       10        20        30        40        50        60        70
          '        '         '         '         '         '         '         '
WMVZ9     MSMKYLML-LFA-AMIIRSFADSGNAIETTSPEITNATTDIPAIRLCGPEGDGYCLHGDCIHARDIDGMY
          :      : :::            :     :     :         :       : :: :      :
WFHU1     MVPSAGQLALFALGIVLAACQALENSTSPLSADPPVAAAVVSHFNDCPDSHTQFCFHGTCRFLVQEDKPA

         71       80        90       100       110       120       130       140
          '        '         '         '         '         '         '         '
WMVZ9     CRCSHGYTGIRCQHVVLVDY-QRSENPNTTTSYIPSPGIMLVLVGI--IIITCCLLS---------VYRF
          : :  :: : :: :  :      :      :         :    :    : ::              :
WFHU1     CVCHSGYVGARCEHADLLAVVAASQKKQAITALVVVSIVALAVLIITCVLIHCCQVRKHCEWCRALICRH

        141    150       160
          '        '         '
WMVZ9     TRRTKLPI------QDMVVP
               :            :
WFHU1     EKPSALLKGRTACCHSETVV

Total score = 1010, 6 breaks
33 identities out of 140 possible matches between residues
300 random runs
Alignment score =     7.64 SD   Standard deviation =     17.86    Mean =  873.55
```

FIGURE 3. Alignment from program ALIGN of vaccinia virus 19-kdalton protein and human transforming growth factor Type 1 precursor. For distantly related sequences, there can be several alignments that obtain the same best alignment score. Moreover, varying the bias and gap penalty will also produce somewhat different alignments and scores.

the past used the Z values to attribute statistical significance and to estimate probabilities that sequences are related, we now believe that it is not appropriate to infer biological significance solely on the basis of measures of statistical significance.

The user must specify a scoring matrix and parameters to control the insertion of gaps in the aligned sequences. We have compared the use of different scoring matrices in this and other programs and found that the mutation data matrix generally is the most sensitive for detecting distant relationships.[27] To obtain reasonable alignments, two types of gap penalty are needed: one is a penalty applied every time a break is inserted in either sequence (the break penalty), regardless of the length of the break; the other assesses a penalty based on the size of the break. In our system this is accomplished by adding a premium (the matrix bias) for each pair of aligned residues; this can be shown to be equivalent to subtracting a penalty for each unaligned amino acid.[36]

The RELATE program, which is similar to one developed independently by Fitch,[37] compares all segments of a given length from one sequence with all segments of the same length from the second sequence. A segment score is calculated for each comparison. This score is the sum of the scores for each individual pair of amino acids occupying corresponding positions within the two segments. If the proteins are related, there will be two populations of scores: a large population from unrelated segments and a very small population from the related segments. The scores from the latter will be higher than those from the unrelated segments and may be detectable as a hump or an elongation of the upper tail of the distribution. The program calculates the expected number of scores in this second population and computes the average of that same number of top scores from the upper tail of the distribution. The sequences are permuted and a distribution of average top scores is generated. The

RELATE score (in SD units) is calculated from the mean and standard deviation of this distribution analogously to the alignment score of program ALIGN. We have found that proteins known to be related obtain ALIGN and RELATE scores above 5 SD. Very distantly related sequences may give scores of 3 to 5 SD or lower. On the other hand, powerful database searching algorithms, such as FASTP, frequently locate fairly long stretches (over 100 amino acids) from unrelated proteins that will give scores in the range of 3 to 8 SD when these selected segments are tested using ALIGN or RELATE. Therefore, we assume that scores within that range indicate evolutionary relatedness only if supported by other biological evidence such as similarity of function, conservation of essential amino acids, and similar secondary and tertiary structures.

The ALIGN program functions best when the sequences being compared are similar along their entire lengths. If the proteins are of markedly different lengths and contain related segments, this program often fails to produce a reasonable alignment and score; in some cases, using a high gap penalty may produce satisfactory results. The RELATE program represents a different approach for assessing the similarity between sequences that cannot easily be aligned. If sequences containing duplicated regions are compared, however, the score from program RELATE can be much higher than would be expected from the degree of overall similarity between them because a segment that matches well to one duplicated region will also match to the others, thus contributing a number of high scores rather than only one.

Testing only selected portions of the two sequences with these programs may also produce artificially high scores. While this procedure can be legitimate in the case of proteins with well-established functional domains, it is often misused. If, for instance, a 30-residue segment is used to search the entire database, it will typically obtain an alignment score of 5 SD when compared with the highest-scoring 30-residue segment found in an unrelated sequence. This is not surprising when you consider that about half a million segments are compared in the search. As mentioned above, FASTP will find even longer regions of similarity between pairs of sequences for which there is no reason to suspect relationship.[32] Inasmuch as there are a fairly limited number of motifs found in the secondary and tertiary structures of proteins,[17] these regions may exhibit some structural similarity. Therefore, we recommend testing for homology over entire sequences or, in some cases, over segments that may reasonably be assumed to represent functional domains. That is not to imply that shorter regions of unusual similarity cannot be used as qualitative evidence of possible distant relationship, but rather to caution against assigning probabilities based on comparisons of short segments with algorithms such as ALIGN.

Although the programs RELATE and ALIGN can provide evidence that two sequences are related, they do not provide a measure of how close the relationship is. In general, it is not valid to assume that two sequences that score 8.5 SD are more closely related than two that score 7.5 SD. As shown in Table 2, it takes many permutations to obtain reproducible SD scores, whereas an SD score based on 100 permutations is frequently high enough or low enough to allow us to accept or reject a hypothesis. Other factors, such as the length of the aligned sequences and the disparity in length between the sequences, also affect the scores obtained with these methods. Consequently, our estimates of how closely sequences are related are based on percentage differences between aligned sequences (see Section VI).

2. *Graphic Matrix Plots and Other Plots of Sequence Characteristics*

Dot matrix graphic plots can be used to locate regions of similarity between two sequences, especially when the lengths are different, and to detect regions of repetitive structure within a sequence or between two sequences. This method will detect all regions having above a certain specified level of similarity, not only the best one as is the case for the FASTP search. It is much quicker than ALIGN or RELATE in revealing the relative locations of

Table 2
REPRODUCIBILITY OF ALIGN SCORES IN INDEPENDENT COMPARISONS OF TWO SEQUENCES

Number of permutations	ALIGN scores Run 1	Run 2
10	17.53	20.35
30	19.35	25.93
100	17.77	18.47
300	17.53	18.83
1000	18.72	18.30
3000	18.12	18.45

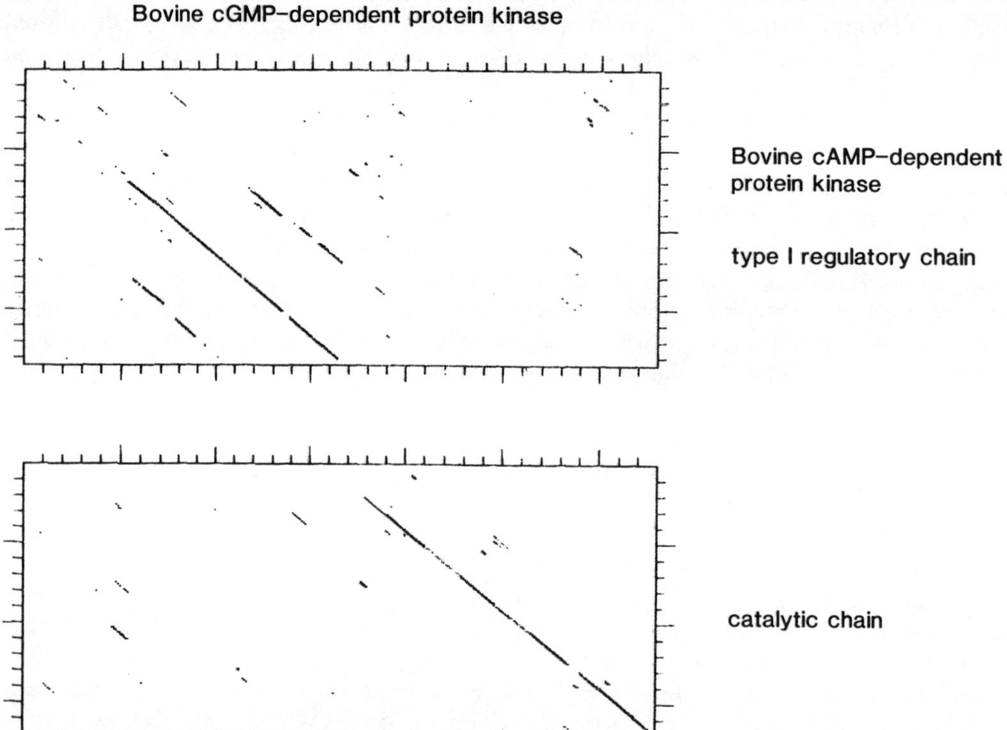

FIGURE 4. Graphic matrix plots comparing the single-chain cGMP-dependent protein kinase with the regulatory and catalytic chains of cAMP-dependent protein kinase. In the upper plot, two off-main diagonal lines represent duplicated regions in the sequences. These plots were generated using the mutation data scoring matrix, a window of 25, and a minimum score of 25.

similar regions in two sequences, especially if one sequence is much longer, if the region in one sequence is interrupted, or if two regions are in an inverted orientation (in one sequence with respect to the other).

Our DOTMATRIX program[38] produces a graphic matrix display (Figure 4) using an algorithm similar to that of program RELATE to find regions of similarity between two sequences. It is similar to the programs described by Maizel and Lenk[39] and Staden.[40] Scores are accumulated from the comparison of all segments of a specified length from one sequence (represented on the horizontal axis) with those of the second sequence (on the vertical axis), not allowing gaps. Each segment comparison score is the sum of all residue-pair scores of the two aligned segments. The pair scores are specified by an appropriate scoring matrix, e.g., a unitary matrix, a hydropathy matrix, or our mutation data matrix. All segment comparison scores greater than or equal to a specified minimum are plotted as a dot at the coordinates corresponding to the midsegment positions. Diagonals in the plot indicate regions of sequence similarity. An offset in a diagonal represents a break in one sequence with respect to the other in the region of similarity. Shorter diagonals above and/or below a main diagonal may represent repeats in one or both sequences; in some cases there are clusters of multiple short diagonals.

A program based on the algorithm of Hopp and Woods[41] may be used to examine sequences for possible similarities in secondary structure or functional regions. In this program, user-specified amino acid scoring data are averaged over each segment of a user-specified length (window size) and the average values are plotted vs. sequence position. The average scores are plotted at the position corresponding to the middle of each segment. If an even window size is specified, the scores are plotted at the position following the midpoint. The program utilizes different sets of amino acid scoring values, including the hydrophilicity data of Hopp and Woods,[41] the hydropathy index of Kyte and Doolittle,[42] and, for indications of possible alpha-helices, beta-sheets, and beta-turns, the data of Chou and Fasman.[43] Alternatively, users may devise their own scoring values.

Figure 5 shows hydrophilicity plots of three proteins that are thought to be distantly related:[44-48] transforming growth factor Type 1 precursor, a portion of epidermal growth factor precursor containing the EGF peptide and the putative membrane-spanning segment, and the vaccinia virus 19-kdalton protein. A prominent feature of all three plots is a markedly hydrophobic region that has the typical appearance of membrane-spanning segments in other proteins.

V. SEARCHING FOR RELATIONSHIPS: EXAMPLES

A. Unexpected Near Identity of Sequences

There are a number of situations in which the near identity of a known sequence with a newly characterized protein is likely to be discovered. These include the finding of the cellular homolog of a viral oncogene (or vice versa), the isolation of a protein from an unexpected source (e.g., from a tissue in which it was not previously known to occur), the discovery of the sequence of an active peptide within that of a larger molecule, or the identification of open reading frames in nucleotide sequences with known gene products. Such identifications can be made with very little sequence information; e.g., fewer than ten accurately known but not necessarily contiguous amino acids. Of course, as the database of known sequences grows, the number of amino acids needed for an identification will also increase, but it will remain surprisingly small for some time to come. This phenomenon is, of course, closely related to the successful isolation of a gene using a probe that represents the coding region for a small segment of the corresponding polypeptide sequence.

1. v-sis and Platelet-Derived Growth Factor

When the nucleotide sequence of the genome of simian sarcoma virus was determined,[49] an oncogenic protein, v-*sis*, of 226 amino acids was predicted. Its sequence was not, however, similar to that of any known protein. Subsequently, human platelet-derived growth factor (PDGF) was partially sequenced[50,51] and the peptides were observed to be very similar to regions of the v-*sis* protein.[1,51] Further investigations[2] revealed that the v-*sis* gene product

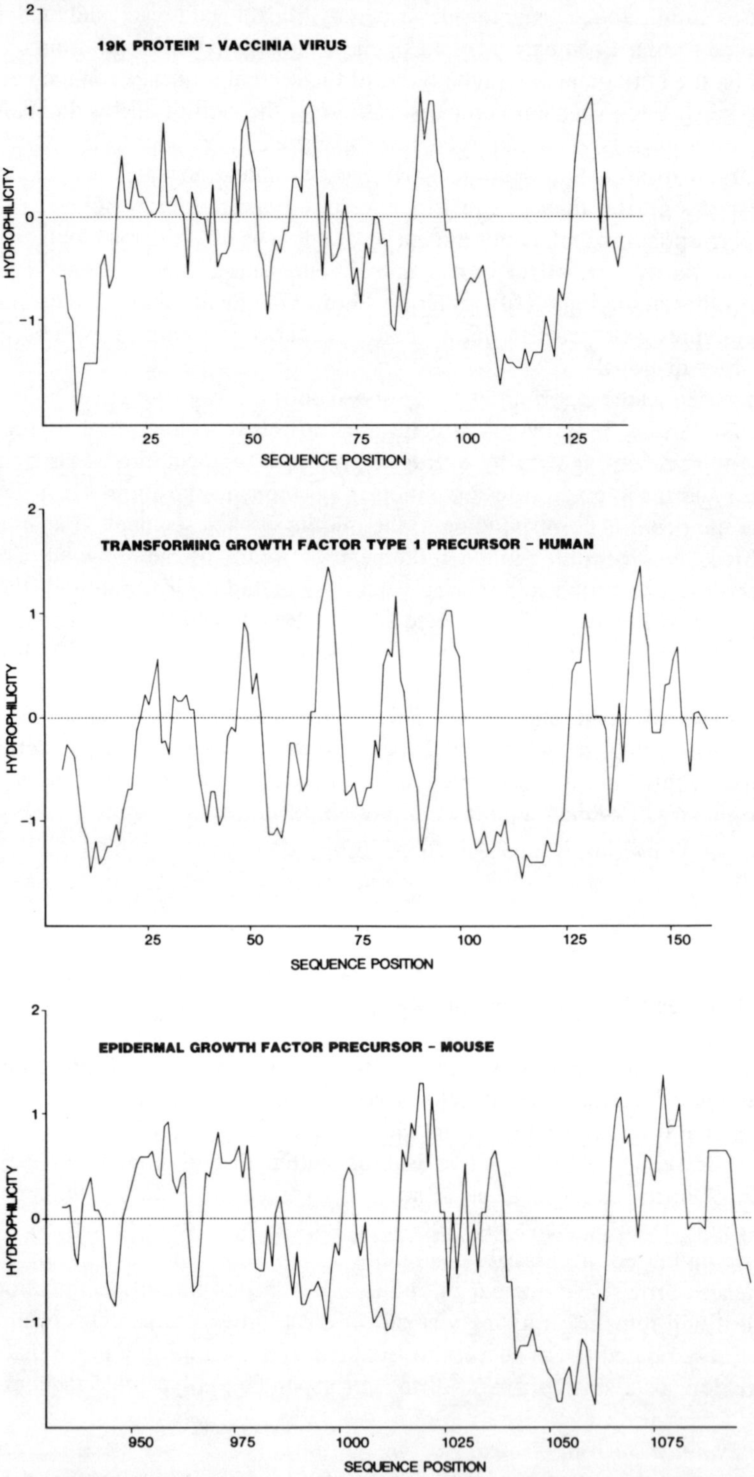

FIGURE 5. Hydrophilicity of human transforming growth factor Type 1 precursor, residues 930-1100 of mouse epidermal growth factor precursor, and vaccinia virus 19-kdalton protein. The data set of Hopp and Woods[41] was used, with a window size of six.

is processed to form a dimer with structural and immunological properties very similar to those of biologically active PDGF. A human cDNA homologous to v-*sis* has been identified and sequenced;[52] it is thought to code for one of two similar chains in the normal PDGF molecule. The previously sequenced peptides accounted for a span of 105 amino acids flanked by arginine residues; except for two discrepancies and six residues not determined, the peptide sequences are identical with this region of the predicted protein sequence.

2. v-erbB and Epidermal Growth Factor Receptor

An oncogenic protein, v-*erbB,* containing 604 amino acids, was predicted from the genomic sequence of avian erythroblastosis virus.[53] The v-*erbB* protein has three domains: an amino-terminal domain of 140 residues, which is probably glycosylated; a mid domain (residues 141-390), the sequence of which is similar to the tyrosine-specific protein kinase family of retroviral oncogene products; and a carboxyl-terminal domain, which may play an important, yet unknown, role in the induction of erythroblastosis, but not of fibrosarcomas, in infected birds. Subsequently, Downward and co-workers[3] sequenced six peptides totaling 83 residues from the human epidermal growth factor (EGF) receptor, a protein known to have tyrosine-specific kinase activity associated with its intracellular domain. In all, 74 residues were found to be identical with regions of the viral *erbB* sequence, a very surprising degree of identity considering the presumed evolutionary distance between the organisms and the failure to detect any protein kinase activity associated with the *erbB* protein. The remarkable resemblance was confirmed when the amino acid sequence of the EGF receptor was deduced from the corresponding nucleotide sequence.[4] In fact, over a stretch of 327 amino acids (residues 89-415 of *erbB* and 669-995 of EGF receptor), the two predicted sequences have only ten differences. The paradox posed by the lack of functional relationship between such structurally similar proteins prompted further attempts, this time successful,[54,55] to demonstrate that the *v-erbB* protein is in fact a tyrosine-specific protein kinase. Inasmuch as the *v-erbB* protein entirely lacks any region related to the EGF-binding domain that controls the kinase activity of the EGF receptor, it is probable that the activity of the *v-erbB* protein is not subject to regulation.

3. v-fgr Protein: A Hybrid Containing a Fragment of Actin

Viral oncogenes can also include nucleotide sequences originating from the host and having no obvious functional significance. Such a case is the v-*fgr* oncogene of the Gardner-Rasheed feline sarcoma virus.[8] This gene, which is fused with the viral *gag* gene, encodes 129 amino acids that are almost identical with the amino end of cytoplasmic actin, followed by 388 amino acids that are 80% identical with the v-*yes* oncogene product of avian sarcoma virus Y73,[56] a protein of the tyrosine-specific protein kinase family. Thus, it is possible for a viral oncogene to arise from recombinational events involving two different cellular genes. It is not known if the actin-like portion of the v-*fgr* protein is in any way involved in its oncogenic activity.

4. Ubiquitin

Ubiquitin has been demonstrated by radioimmunoassay in a variety of mammalian tissues and in many other organisms; it is thought to be present in all living cells. This 76-residue sequence is very highly conserved, as human and fruit fly sequences are identical.[57,58] The wide distribution and conserved sequence suggest that ubiquitin has an ancient, important function in all cells. When a protein is so widespread, it is likely to be isolated under different circumstances without recognizing its identity. For example, ubiquitin has also been called thymosin polypeptide beta$_1$, protein S in trout testis, nonhistone chromosomal protein HMG-20, and ATP-dependent proteolysis Factor I.[59] When an amino-terminal sequence of nuclear protein A24 was published,[60] one of us discovered it to be identical with that of ubiquitin.[61]

Nuclear protein A24 occurs in mammalian tissues in nuclear and nucleolar chromatin, but not in ribosomes, and was found to be a complex of two polypeptide chains: histone H2A and ubiquitin.[62]

On another occasion, the protein sequence database was searched to identify a partially sequenced protein isolated from pituitary tissue.[63,64] Lys, Arg, or Met had been identified at 8 of the first 33 positions; nothing was known about the sequence at the other positions except that these three amino acids did not occur. Only one sequence, ubiquitin, matched the known information perfectly. As a consequence of this search, Dayhoff and Orcutt suggested a strategy to optimize the probability of identifying radiolabeled proteins.[64] A tissue would be incubated with a mixture of all necessary amino acids under conditions that permit protein synthesis. Several of the amino acids would carry one radioactive label, several another label, and the rest would be unlabeled. Proteins synthesized by the tissue can be separated, the label carried at each of the first 30 or so positions in a sequence can be determined, and the database searched for sequences that match the pattern. With such data, most human proteins could be identified if the sequences were in the database.

5. Gamma-Trace: A Protein in Search of a Function

Human gamma-trace was described as a low molecular weight basic protein found in small amounts in cerebrospinal fluid, plasma, saliva, and semen. Later it was demonstrated also in cells in the brain, adrenal medulla, pancreas, and pituitary. On the basis of its location and some sequence similarities with certain hormones, it was suggested to be part of the peptidergic gastroenteropancreatic neuroendocrine system.[65] When one of several small acidic proteins isolated from saliva was sequenced,[66] it proved to have more than 50% identity with gamma-trace. The function of this protein, called SAP-1, was also not known. Meanwhile, two groups sequenced cystatin, a thiol proteinase inhibitor isolated from chicken egg white.[67,68] One of these groups subsequently isolated a similar inhibitor from sera of patients with autoimmune disease and showed that its amino acid composition and the sequence of its first 50 residues were identical with those of gamma-trace.[69] After these results were reported, SAP-1 was also demonstrated to have thiol proteinase activity.[70]

B. Distantly Related Proteins

1. Some Definitions

We have for many years used the word ''homologous'' to describe proteins or genes that share a common evolutionary ancestor. The inference of common evolutionary ancestry is often considered obvious if sequences of a reasonable length, when aligned with very few internal gaps allowed, have perhaps 30% identical amino acids. Homologous proteins may be encoded by the same gene in different species or by genes that have been transferred between species, or, within a species, by genes that originated from duplication of an ancestral gene. In this usage, proteins or genes are either homologous or not homologous; the term ''percent homology'' has no meaning. The word ''homology'' is also very often used to indicate an observed similarity, without reference to the etiology of the similarity. Sequence similarity can be measured in terms of percentage of identical amino acids or percentage of similar amino acids. Authors vary in their choice of what constitutes a similarity. Some authors simply declare an observed similarity to be unusual, others apply some criteria based upon computer comparisons with many sequences or with scrambled sequences. Similarity of sequences or of regions of sequences is often interpreted to indicate evolutionary relationship, common function, similar two- or three-dimensional structure, or antigenic cross reactivity. Such conclusions about the biological significance of observed similarities cannot be proved solely by sequence comparison methods; however, these methods can identify proteins with sufficient sequence similarity to warrant an investigation into their possible structural, functional, or evolutionary relationships.

Table 3
ALIGNMENT SCORES OF EUKARYOTE PROTEIN KINASES AND RELATED SEQUENCES

	Receptors (human) EGF	Receptors (human) Insulin	c-*src* (chicken)	c-*src* (fruit fly)	c-*mos* (human)	CDC28 prot. (yeast)	Phos. b kinase (rabbit)	Protein kinases (bovine) cGMP	Protein kinases (bovine) cAMP
EGF rec.	—	23.7	22.1	23.0	9.6	12.6	7.7	10.1	10.4
Ins. rec.	23.7	—	22.3	22.0	13.5	11.7	9.6	10.4	8.7
c-*src*, chicken	22.1	22.3	—	38.5	12.4	10.7	9.6	9.9	12.4
c-*src*, fly	23.0	22.0	38.5	—	10.6	10.3	7.9	11.4	10.3
c-*mos*	9.6	13.5	12.4	10.6	—	9.2	5.1	10.1	8.3
CDC28	12.6	11.7	10.7	10.3	9.2	—	10.2	9.9	10.9
Phos. b kinase	7.7	9.6	9.6	7.9	5.1	10.2	—	16.0	19.9
cGMP PK	10.1	10.4	9.9	11.4	10.1	9.9	16.0	—	30.4
cAMP PK	10.4	8.7	12.4	10.3	8.3	10.9	19.9	30.4	—

2. Oncogenes, Protein Kinases, and Growth Factor Receptors

In 1982 we reported[6] the discovery that the carboxyl-terminal half of the retroviral *src* gene product is distantly related to the catalytic chain of mammalian cAMP-dependent protein kinase. This was the first demonstration of a relationship between oncogenic proteins and important cellular regulatory proteins, and it supported the idea that viral oncogenes originated from the host genome and encode products that are closely related to cellular proteins essential for normal growth and development. Moreover, because the *src* protein is a tyrosine-specific protein kinase whereas the mammalian enzyme phosphorylates serine and threonine, it provided the first evidence for structural relationships between protein kinases that respond to different regulatory signals and serve different regulatory roles.

The hypothesis that the *src* and *mos* gene products and the catalytic chain of the cAMP-dependent protein kinase are related was based on the finding of several segments with unusually high scores using program SEARCH. The sequences were then tested using program RELATE, which gave scores ranging from 3.1 to 7.8 SD. It was possible to construct an alignment spanning 250 positions, with 21 to 24% identities between any two of the sequences. The *src* transforming protein was known to have protein kinase activity, which was localized in the same region of the molecule that could be aligned with the mammalian protein kinase. Moreover, the viral sequences conserved the lysine residue proposed to be the ATP-binding site in the mammalian enzyme. On the basis of this evidence, we proposed that the cellular homologs of these viral transforming proteins are members of a superfamily of distantly related protein kinases that are normal constituents of mammalian cells. Since that time several cellular proteins have been shown to have domains homologous to the ATP-binding and catalytic domain of cAMP-dependent protein kinase, including cGMP-dependent protein kinase,[71] the gamma chain of the calcium-regulated phosphorylase b kinase,[72] and the cytoplasmic domains of epidermal growth factor receptor (which has tyrosine-specific kinase activity[4]) and of insulin receptor.[5] An ever-increasing number of retrovirally encoded transforming proteins and their cellular homologs[10] and a yeast cell division control protein[73] have also been shown to belong to this superfamily of proteins. Table 3 shows alignment scores of the homologous domains of a selection of the eukaryote proteins. In Section VI.E below, another approach to the analysis of these relationships is shown.

The cAMP-dependent protein kinase is a tetrameric molecule consisting of two regulatory chains and two catalytic chains. The regulatory chains have three identifiable domains: an amino-terminal domain including the sites for dimerization and phosphorylation and two homologous nucleotide-binding domains.[74] These are also homologous with the nucleotide-

LIGAND-BINDING DOMAIN

```
                                           1                   2                   3                   4
                           1 2 3 4 5 6 7 8 9 0 1 2 3 4 5 6 7 8 9 0 1 2 3 4 5 6 7 8 9 0 1 2 3 4 5 6 7 8 9 0 1 2 3 4 5

Repeat I: 23-63            /V G D R C - E R N E F Q C Q D - - G K C I S Y K W V C D G S A E C Q D G S D E S Q E T C/
Repeat II: 64-104          /L S V T C - K S G D F S C G G R V N R C I P Q F W R C D G Q V D C D N G S D E Q G - - C/
Repeat III: 105-143        /P P K T C - S Q D E F R C H D - - G K C I S R Q F V C D S D R D C L D G S D E A S - - C/
Repeat IV: 144-184         /P V L T C - G P A S F Q C N S - - S T C I P Q L W A C D N D P D C E D G S D E W P Q R C/
Repeat V: 193-231          /D S S P C - S A F E F H C L S - - G E C I H S S W R C D G G P D C K D K S D E E N - - C/
Repeat VI: 232-270         /A V A T C - R P D E F Q C S D - - G N C I H G S R Q C D R E Y D C K D M S D E V G - - C/
Repeat VII: 271-313      /V N V T L C E G P N K F K C H S - - G E C I T L D K V C N M A R D C R D W S D E P I K E C/
```

TYPE B EGF-LIKE REPEATS

```
                                         1                   2                   3                   4
                       1 2 3 4 5 6 7 8 9 0 1 2 3 4 5 6 7 8 9 0 1 2 3 4 5 6 7 8 9 0 1 2 3 4 5 6 7 8 9 0 1 2 3 4 5

Repeat A: 318-352 /C L D - - - N N G G C S H V C - - - - - - - N D L K I G Y E C L C P D G F Q L V A Q R R - C/
Repeat B: 358-392 /C - Q - - - D P D T C S Q L C - - - - - - - V N L E G G Y K C Q C E E G F Q L D P H T K A C/
Repeat C: 667-711 /C E R T T L S N G G C Q Y L C L P A P Q I N P H S P K F T C A C P D G M L L A R D M R S C/
```

THIRD REPEAT PATTERN

```
                                         1                   2                   3                   4                   5
                       1 2 3 4 5 6 7 8 9 0 1 2 3 4 5 6 7 8 9 0 1 2 3 4 5 6 7 8 9 0 1 2 3 4 5 6 7 8 9 0 1 2 3 4 5 6 7 8 9 0

Repeat 1: 397-438 /S I A Y L F F T N R - H E V R K M T L D - - - R S E - Y T S L I - - P N - L R N V V A L D T E V A S/
Repeat 2: 439-485 /N R I Y W S D L S Q - R M I C S T Q L D R A H G V S S Y D T V I - S R D - I Q A P D G L A V D W I H/
Repeat 3: 486-528 /S N I Y W T D S V L - G T V S V A D T K - - - G V K - R K T L F - R E N - G S K P R A I V V D P V H/
Repeat 4: 529-572 /G F M Y W T D W G T P A K I K K G G L N - - - G V D - I Y S L V - T E N - I Q W P N G I T L D L L S/
Repeat 5: 573-615 /G R L Y W V D S K L - H S I S S I D V N - - - G G N - R K T I L E D E K R L A H P F S L A V F E/
Repeat 6: 616-657 /D K V F W T D I I N E A I F S A N R L T - - - G S D - V N - L L - A E N - L L S P E D M - V L F H N/
```

FIGURE 6. Repeating patterns of the human LDL receptor sequence.

binding domain of the *Escherichia coli* catabolite gene activator protein, also called the cAMP receptor protein.[75] The carboxyl-terminal domain of the latter protein binds to DNA and is unrelated to the protein kinase regulatory chains. As illustrated in Figure 4, the cGMP-dependent protein kinase contains within a single chain sequences homologous with both regulatory and catalytic chains of cAMP-dependent protein kinase.[71] Thus, it is apparent that a single protein chain can consist of several domains with different evolutionary origins and histories.

3. *Domains of LDL Receptor Protein Found in Other Proteins*

Nucleic acid sequencing has allowed the prediction of the complete sequences of large proteins, membrane-bound proteins, and proteins present in low concentrations. The LDL receptor protein is an example of a large multifunctional protein; it is composed of a signal sequence, a ligand-binding domain, a domain with homology to regions of the epidermal growth factor (EGF) precursor, a domain rich in serine and threonine residues that bind carbohydrate, a transmembrane domain, and a cytoplasmic domain.[76] The ligand-binding domain consists of a sevenfold repeat of a 40-residue pattern (see Figure 6), a single copy of which has been found also in the sequence of complement factor C9.[77,78] Analysis of the LDL receptor gene revealed that four of these repeats are encoded by separate exons whereas the other three occur in a single exon, suggesting that introns once occurred between them but have been lost.[77] The functional significance of the homology between C9 and LDL receptor regions is a matter of speculation; however, it is known that LDL can inhibit the polymerization of C9.[79]

Following the ligand-binding domain is a region encoded by eight exons that shows strong homology with residues 400 to 790 (approximately) of mouse EGF precursor[76] and with the corresponding region of the human EGF precursor,[80] which also is encoded by eight exons. Five of the exon-intron junctions in this region occur at exactly the same positions in the two genes and three others are slightly displaced.[80] Most of this region of the EGF precursor shows intrasequence homology with its amino-terminal region (approximately residues 21 to 365),[81] and the similarities between this portion of the EGF precursor and the homologous domain in the LDL receptor are also quite striking. These regions are also very repetitious, containing two different patterns. One pattern is found three times in the LDL receptor[77] (Figure 6), five or six times in the EGF precursor, and one time each in coagulation Factors IX and X and protein C.[81] The second type of pattern, while also about 40 residues long, is less well defined. No single amino acid is absolutely conserved, but there are areas where a consensus sequence can be derived. There are six tandem occurrences of this pattern in the LDL receptor (Figure 6) and in each of the homologous regions of the EGF precursor.

The EGF precursor has several copies of each of two types of EGF-like sequences.[81] Type A sequences are most similar to EGF itself; the regions found in the LDL receptor are Type B. Both types typically contain six cysteine residues thought to form three disulfide bonds.

Thus, the large LDL receptor protein is composed of many rather short (20 to 50 amino acids) segments: signal sequence (21 residues), seven 40-residue repeats (the ligand-binding domain), three Type B EGF-like regions (35-45 residues each), six repeats of a different 40-residue pattern; the remainder of the chain, about 150 residues, contains the highly glycosylated region, the transmembrane region, and the intracellular region. We must consider the possibility that each of these different types of segments had an independent evolutionary origin. Unfortunately, the small size of these segments makes it very difficult to distinguish between common evolutionary origin, convergent evolution of sequence similarity because of similar function, or chance occurrence of sequence similarity.

4. Serine Proteases of the Coagulation and Complement Systems

In the human body, as well as in organisms as diverse as bacteria and invertebrates, are found a number of proteolytic enzymes with a catalytic mechanism similar to the digestive enzymes trypsin and chymotrypsin. These proteases also show sequence homology, which is particularly evident in the region of the catalytic serine residue. Moreover, the three-dimensional structure of bacterial and mammalian pancreatic enzymes has been shown to be similar even when there is little residual sequence homology.[82,83] A number of proteins of the blood coagulation and complement systems also have serine protease activity and sequence homology with this group. These are typically larger molecules, having the trypsin-like domain at the carboxyl end, preceded by various other domains. The evolution of these functionally diverse proteins has been reviewed recently.[84]

At least six different types of domains have been identified in the mature enzymes (see Table 4). An amino-terminal calcium-binding domain about 45 residues long is found in prothrombin,[85,86] protein C,[87] and Factors IX[88] and X.[89,90] It contains at least ten glutamic acid residues that undergo vitamin K-dependent modification to gamma-carboxyglutamic acid. Factors IX and X, urokinase (plasminogen activator), tissue plasminogen activator (TPA), and protein C all have domains with strong sequence homology to EGF.[81,91-94] This region is known to be encoded by a separate exon in the genes for TPA and Factor IX.[94,95] Similar Type A EGF-like domains occur in the EGF precursor,[81] in transforming growth factor Type 1,[47,48] in the vaccinia virus 19-kdalton protein,[44-46] and in complement factor C9,[95a] none of which has trypsin-like domains (see Figure 7). As mentioned above, three of the coagulation factors also contain Type B EGF-like domains. The kringle domains are about 80 amino acids long and contain a characteristic pattern of six disulfide-bonded cysteine residues. Plasminogen has five kringle regions, prothrombin and TPA each have two, and urokinase has one.[85,86,92-94,96] It is interesting that each kringle region of TPA is encoded by two exons,[94] but a kringle region in human prothrombin is entirely encoded within a single exon.[97] Kringle regions have not so far been found in other proteins. TPA has one region similar to the 55-residue "finger-like" structures that are present in multiple copies in the fibrin-binding domains of fibronectin.[98-100] And, finally, complement factor B contains a threefold repeat of a 60-residue pattern found also in beta$_2$-glycoprotein I and in certain nonenzymic complement factors.[101-105]

Table 4
DOMAINS OF COAGULATION AND COMPLEMENT FACTORS[a]

	Calcium-binding	Finger	EGF-like Type A	EGF-like Type B	Kringle	$Beta_2$-glycoprotein I	Trypsin
Plasminogen (human)	—	—	—	—	83-161 165-242 255-332 357-434 461-540	—	561-790
Prothrombin (human)	1-45	—	—	—	65-143 170-248	—	323-581
Factor IX precursor (human)	47-92	—	97-128	134-170	—	—	227-462
Factor X (human)	1-45	—	50-81	89-124	—	—	195-428
Urokinase (human)	—	—	11-42	—	50-131	—	(158-410)
TPA precursor (human)	—	39-84	86-119	—	127-208 215-296	—	311-562
Factor B (human)	—	—	—	—	—	9-74 75-134 137-194	460-736
Protein C (bovine)	1-45	—	50-89	98-133	—	—	170-408

[a] Boundaries are approximate; EGF-like and kringle boundaries are from the first Cys to the last Cys of the characteristic pattern.

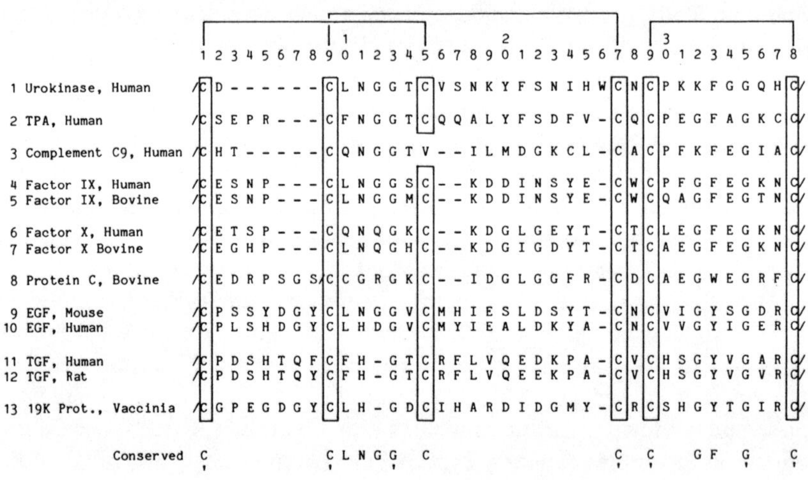

FIGURE 7. Alignment of Type A EGF-like domains in various proteins.

The occurrence of particular functional domains in some members of a group of related proteins and not in others and in proteins that are otherwise unrelated implies that we must in these cases analyze the relationships and evolutionary histories of the domains rather than of the entire protein sequences. Indeed, the shuffling of exons among genes is a mechanism for the extremely rapid evolution of new proteins,[106] especially those that function in complex systems in which they interact with several other molecules.

VI. EVOLUTIONARY RELATIONSHIPS AND TREES

A. Computer Derivation of Trees

A dendrogram, or tree, can be used to display graphically the degree of difference among a group of protein sequences. If the proteins, or segments of the proteins, are assumed to be related, then the dendrogram becomes a tree representing evolutionary events, such as species divergence points, gene duplications, etc. These events have occurred over a certain period of time; however, it is not usually possible to assign an absolute time scale to the events depicted on the tree. To do so would require knowledge of the rate of mutational acceptance, which in general is not constant. Furthermore, sequence comparisons yield only the relative number of differences or the distance between sequences. They do not provide any information concerning the placement of the point of earliest time on the base of the tree. Thus, the temporal order of the divergence points on the tree cannot be absolutely determined. Several different kinds of trees, of varying levels of complexity, can be distinguished depending on the kinds of information that are included, such as several kinds of related proteins from the same organism, or one kind of molecule from a number of species, or more than one protein from more than one species. When multiple protein forms are related, it can be a problem to distinguish species divergences and sequence divergences.

To obtain a tree, one first constructs an alignment of the sequences. This is generally done by a combination of computer methods[16,31,35] and biological reasoning. Although computer methods will produce a ''best'' alignment of any two sequences under a given set of parameters, this does not guarantee the most biologically realistic alignment; furthermore, the alignments generated for different pairs of sequences may be mutually inconsistent. The alignment must be adjusted to reflect biological considerations, i.e., the proper alignment of active sites, etc. From the alignment is derived a matrix of percentage differences; these values are corrected for inferred parallel and superimposed mutations to yield a matrix of evolutionary distances between sequences, which is used to derive evolutionary trees.[107]

Our method for deriving trees[16,107,108] is a modification of the matrix method first described by Fitch and Margoliash;[109] trees are constructed from a matrix of the total estimated evolutionary change between each pair of sequences. For each possible topology, a set of branch lengths is determined that gives the best weighted least-squares fit of the terms of the reconstructed and original matrices. The weights are inversely proportional to the variance of each matrix element (a function of its size). For trees with less than nine branches, all possible topologies are tried. Each of these branches can represent a single sequence, several sequences for which a topology is specified, or a group of closely related sequences (the average matrix elements are used to represent this group). A tree with nine or more branches can be approached by running various combinations of branches until a tentative picture of the overall topology emerges. One can then specify the entire topology and try to improve the overall length of the tree by allowing individual sequences or branches of specified topology to move.

The best topology is usually the one with minimal total (absolute value) branch length, all (or nearly all) positive branches, and consistency with biological data. The branch lengths are expressed in accepted point mutations per 100 residues (PAMs).[16] The location of the point of earliest time (i.e., the connection of the trunk to the branching structure) cannot be determined directly from the algorithm but must be estimated from other considerations. In principle, a sequence known to branch off first from all of the others could provide the trunk for the rest of the tree. In practice, however, sequences that diverged early are often so different that they cannot be placed relative to the remaining topology with certainty.

We have compared the precision of the least-squares matrix method (described above) with that of a maximum parsimony method, based on generating ancestral sequences from an alignment of related sequences,[110] for correctly reconstructing sparse trees of distantly related sequences. From an initial 100-residue sequence of average amino acid composition, a set of other nodal sequences and terminal sequences with predetermined numbers of "mutations" was generated by random processes. The results for trees containing four sequences and having branches of equal length (L) showed that, for distant sequences ($L > 50$), the matrix method is clearly superior to the ancestral sequence method. It also had the advantage of not being as sensitive to the details of the alignment as is the ancestral sequence method. Peacock and Boulter[111] also found the matrix method to be more accurate for sequences with greater than 30 amino acid substitutions per 100 residues. Astolfi et al.[112] have compared seven methods of reconstructing evolutionary trees: least squares, cluster analysis, single linkage, average linkage, maximum likelihood, spectral analysis, and minimum average covariance. They found the least squares, average linkage, and maximum likelihood methods to be best. Each of these had some advantage over the others for a particular purpose, so the choice is a matter of user preference. More recently, Blanken et al.[113] compared eight methods of constructing evolutionary trees from sequence data, including the method we use. For simulations involving branches with unequal evolutionary rates, only a new method proposed by them performed significantly better than several other methods including ours. Their new method, however, involves using a set of several sequences of known phylogeny. A variation of their method that requires no *a priori* phylogenetic information was among the methods that found the correct topology 67 to 69% of the time.

We have also used sets of computer-generated sequences to test criteria for determining the best topology. Three criteria were examined:

1. The minimum sum of the absolute lengths of all of the branches.
2. The minimum net sum of the lengths of all of the branches.
3. The minimum weighted variance of the reconstructed matrix from the input matrix.

Of the three criteria tested, the reconstructed topology with the minimal sum of absolute branch lengths in every case gave the correct solution more often than (or as often as) the other criteria. Fitch and Smith[36] have also presented arguments, based on general principles, for using this criterion for selecting the "best" evolutionary tree. We also determined in the course of this study, that when we used the matrix of estimated accepted point mutations per 100 residues (PAMs) the correct topology was reconstructed more frequently than when we used a matrix of percentage differences.

B. Uses of Alignments and Trees

Alignments and trees are very useful organizing tools in that they provide pictorial representations of the relative amounts of difference among a group of sequences. Alignments can illustrate regions of primary and secondary structural and functional correspondence among sequences, and positions of active sites, modified residues, disulfide bonds, and other features. Trees can diagram possible genetic events underlying the evolution of a group of proteins and can furnish a convenient way to compare alternative evolutionary possibilities.

We routinely make alignments of protein sequences that are similar or known to be related, and we use the percentage difference to organize the sequences into families and superfamilies.[31,114] We construct new trees and update previously derived trees for any of several purposes:

1. To provide new information for our composite phylogenetic tree
2. To organize a group of sequences known or thought to be related
3. To display evolutionary relationships and to serve as a scheme for further investigation
4. To incorporate new information (e.g., on the secondary or tertiary structure of the proteins) that may necessitate a realignment of the sequences

5. To construct ancestral (nodal) sequences and/or predict the primary structure of undetermined, closely related sequences
6. To study the possible transfer of genetic material within and between various genomes and the origins and evolutionary relationships of genomes

C. Problems in the Interpretation of Evolutionary Trees

There continue to be some general problems, some of which are briefly discussed below, associated with derivation and interpretations of protein alignments and trees:

1. Estimating the point of earliest time on a tree (see above)
2. Distinguishing protein divergences from species divergences
3. Interpreting trees with proteins whose genes are changing at different rates
4. Assigning approximate times that are consistent with the fossil record to events on trees
5. Aligning sequences of different lengths and properly placing gaps
6. Interpreting sequence comparisons when various genetic transfers may be involved
7. Choosing the proper boundaries for domains that may have different evolutionary origins than other parts of the sequence, including similar regions within the sequence that may have arisen by internal gene duplication
8. Dealing with areas of uncertainty in branching order, especially for distantly related sequences

The interpretation of evolutionary trees derived from sequence data has been based on several assumptions that are open to question. The ''molecular clock'' hypothesis assumes that evolution proceeds mainly by point mutations that occur at the same average rate in most loci. Differences in the rate of evolution are due to selection; widely different rates of accumulation of accepted point mutations are seen. Nevertheless, the rate of evolution of the same protein occupying the same physiological niche generally appears to be fairly constant within the span of vertebrate evolution. Although our method does not assume that each protein on the tree is evolving at the same rate, it does assume that amino acids at each position in a particular protein evolve at the same rate, the average for that protein.

Nodes that correspond to gene duplications can be identified on a tree if both proteins encoded by the duplicated loci have been sequenced from the same organism. However, if one does not realize that a gene duplication has occurred, it could be misinterpreted as a species divergence, leading to an erroneous dating of the species divergence or an overestimate of the rate of change of the sequences. After a gene duplication, the rates of change of the two genes may become quite different, one of them perhaps undergoing accelerated change until it settles into a new physiological niche. These rate differences introduce uncertainty into estimates of when the duplication occurred.

It now appears that mechanisms other than the accumulation of point mutations may be common in the evolution of higher organisms. Domain transfer, gene conversion, and minigene insertion have been proposed to explain observed phenomena. Even though the underlying mechanisms are still under investigation, it is evident that there are genetic mechanisms that effectively accelerate or retard the accumulation of differences between duplicated genes. Thus, if a gene duplicated before the divergence of the evolutionary lines leading to two species and if by gene conversion the sequences are kept homogenized for some time within each evolutionary line, it will appear that the original gene duplicated independently in the two lines since their divergence. Figure 8 illustrates the tree topologies that might be found when (1) duplicate genes evolve independently by the accumulation of point mutations and (2) the duplicated genes are homogenized for a significant period of time. In the topologies, diamonds mark the divergences that appear to represent gene duplications. In (2) these divergence points actually represent the point when the duplicated

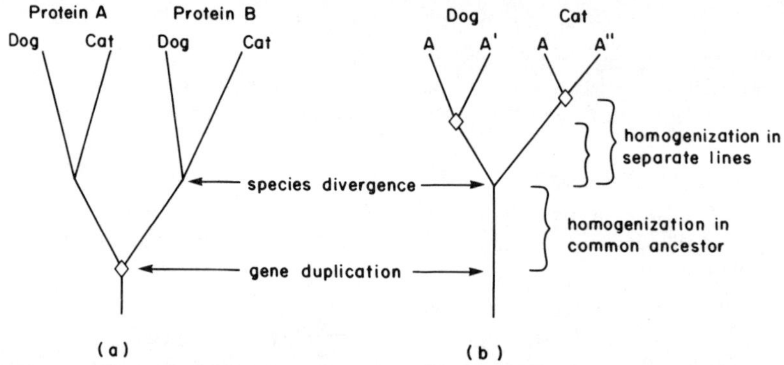

FIGURE 8. Two topologies illustrating the possible effect of gene conversion on the evolution of duplicate genes.

genes began to evolve independently; the original duplication is not evident and it appears instead that independent duplications have occurred since dog and cat diverged. There are a number of examples in which evolutionary trees derived from sequences show what appear to be independent duplications in several lines, e.g., hemoglobin beta and gamma chains in primates and artiodactyls and immunoglobulin gamma chain subclasses in human and mouse. Moreover, other duplications appear to be anomalously recent; a good example is growth hormone and choriomammotropin, which appear to have diverged 50 to 60 million years ago, yet both proteins are characteristic of placental mammals, which diverged about 100 million years ago. Most probably the gene duplications occurred near the time of emergence of placental mammals and more recent gene conversion events account for the similarity between the human proteins.[115] Although it is useful and instructive to derive evolutionary trees from protein and nucleic acid sequence data, these trees must be interpreted with caution, especially in the case of multigene families.

D. Phylogenetic Trees

For more than 20 years, we have been using protein (and nucleic acid) sequences to study the evolution of organisms and to construct a broad outline of phylogeny. This phylogenetic "Tree of Life" is the largest and most complex kind of tree; it depicts relationships among phyla and kingdoms of organisms and their evolution on a geologic time scale. It provides a framework for correlating the evolutionary histories of prokaryotes and their bacteriophages and plasmids, of eukaryotes and their viruses, and of the symbiotic origins of eukaryotes from some prokaryote genomes, and for studying other evolutionary problems. We want to correlate times of biological events on this tree with physical and geological events in the history of the earth.

Most of this tree was constructed from three smaller evolutionary trees derived from three kinds of sequences, two protein trees (c-type cytochromes and ferredoxins) and a nucleic acid tree (5S ribosomal RNA); a simplified form is presented in Figure 9.[116] C-type cytochromes are highly conserved, slowly changing proteins; homologs can be recognized in most living species. Therefore, they are very useful for studying evolution over an extremely long period of time. The phylogenetic tree of c-type cytochromes (Figure 10) contains eukaryotic mitochondrial cytochrome c, bacterial cytochromes c2, c551, and c555, and chloroplast and cyanobacterial cytochromes c6.[117] This tree provides evidence for multiple symbiotic origins of chloroplasts and mitochondria from prokaryotes, as does the tree for plant 2Fe-2S ferredoxin.[118]

Our latest tree of bacterial ferredoxins (Figure 11), which has been recently revised because of evidence for a sequence reversal,[119] shows an ancient ancestral duplication. Evidence of

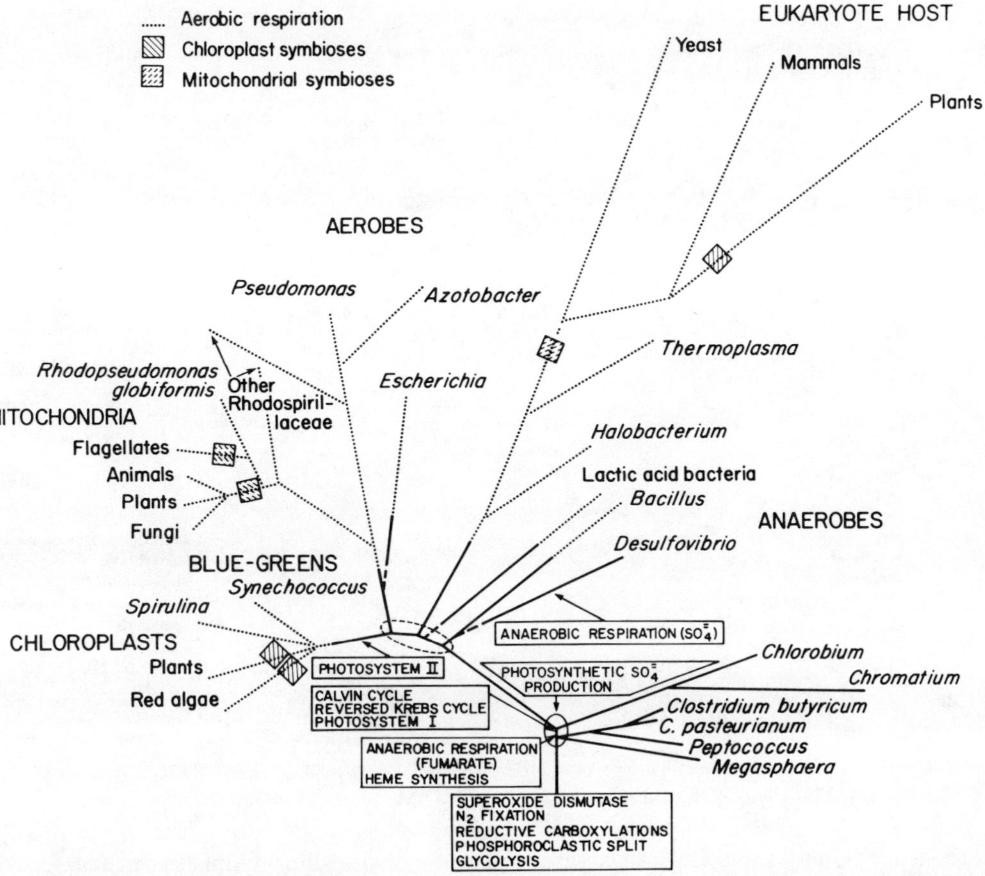

FIGURE 9. The phylogenetic tree of life. (From Barnabas, J., Schwartz, R. M., and Dayhoff, M. O., *Origins Life*, 12, 81, 1982. With permission.)

this duplication is found in all ferredoxins, and allows us to put a base, as the point of earliest time, on this tree and on the composite phylogenetic tree.[108,116,117] The phylogenetic tree of 5S ribosomal RNA (Figure 12) provides evidence for the symbiotic origins of chloroplasts and mitochondria.[120] However, this molecule is less useful for distinguishing events during the early evolution of eukaryote kingdoms.[120] Other protein sequences that provide information about the evolution of chloroplasts and mitochondria from prokaryotes include various chains of the ATPase (EC 3.6.1.34) molecule (e.g., Figure 13) and the large chain of ribulose-bisphosphate carboxylase/oxygenase (EC 4.1.1.39) (Figure 14).

E. Superfamily Trees

Superfamily trees are evolutionary trees of distantly related proteins or protein families. These protein sequences have detectable similarity but are more than 50% different. The relationship is supported by other kinds of evidence, for example, similarity of secondary or tertiary structure or of function. The evolutionary tree of some Type 1 copper-binding proteins and ragweed allergen Ra3 (Figure 15) was derived from an alignment based on X-ray crystallographic data. This tree displays a new evolutionary relationship that we discovered, that between the ragweed protein and the Type 1 copper-binding proteins,[121] and thus adds a family to the superfamily tree.

A second example of a superfamily tree is that of the kinase-related transforming proteins (Figure 16). This tree may be useful for studying the origin and relationship of various viral

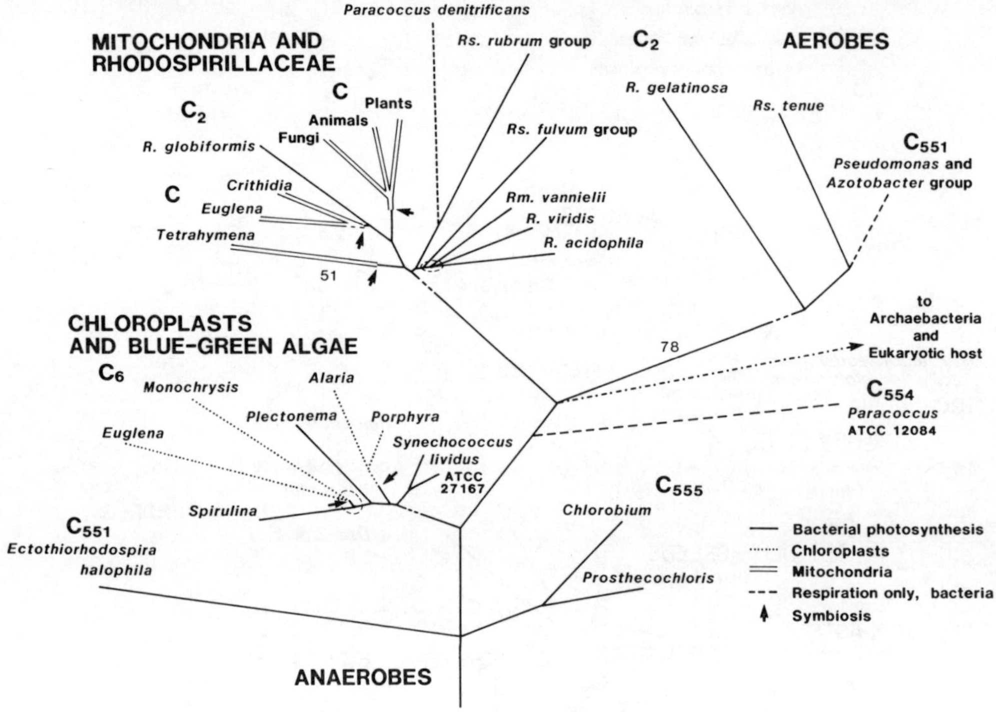

FIGURE 10. Evolutionary tree of c-type cytochromes. (From Hunt, L. T., George, D. G., Barker, W. C., *BioSystems*, 18, 223, 1985. With permission.)

oncogenes (v-*onc*) and cellular oncogenes (c-*onc*), including the possible transfer of genetic material between viruses and eukaryotic hosts and the functional relationships of the different domains within these protein sequences. Some points that can be observed about sequences on this tree include the following: the v-*erbB* protein, for which kinase activity has recently been found,[54,55] and the EGF receptor are much closer than one would expect for sequences from an avian virus and a mammal; the murine v-*abl* (mouse host) and the fruit fly c-*abl* are also closer than would be expected; the feline v-*fes* and v-*fgr* are much further apart than would be expected. This tree illustrates a situation often seen when comparing very distantly related sequences: the (four) main branches are very long and the internodal distance is very short; further, the order of branching is uncertain.

F. Genetic Transfers and Trees of Domains

One of the biggest problems in studying protein evolution is to determine which proteins or parts of proteins are indeed related, so that one is able to compare sequences derived from evolutionarily related genetic material. The extent of transfer of genetic material (genes, parts of genes, or other genetic elements) at all levels — within chromosomes, between chromosomes, between organisms — has been gradually discovered over the past few years. One has to keep such possibilities in mind when deriving and interpreting evolutionary trees; conversely, trees derived from sequences can provide evidence for these transfers. The superfamily tree of the kinase-related transforming proteins discussed earlier (Figure 16) is one example. The tree showing a possible evolutionary scheme for the epidermal growth factor domain (Figure 17) is another example of transfer of a part of a gene that codes for a functional domain, not only among eukaryote genes but also involving a eukaryote virus genome.[44-46] So far this EGF-like domain has been found in normal EGF, in transforming GF Type 1,[47,48,122] in several mammalian blood coagulation factors[81,91,93,94] (in two of these,

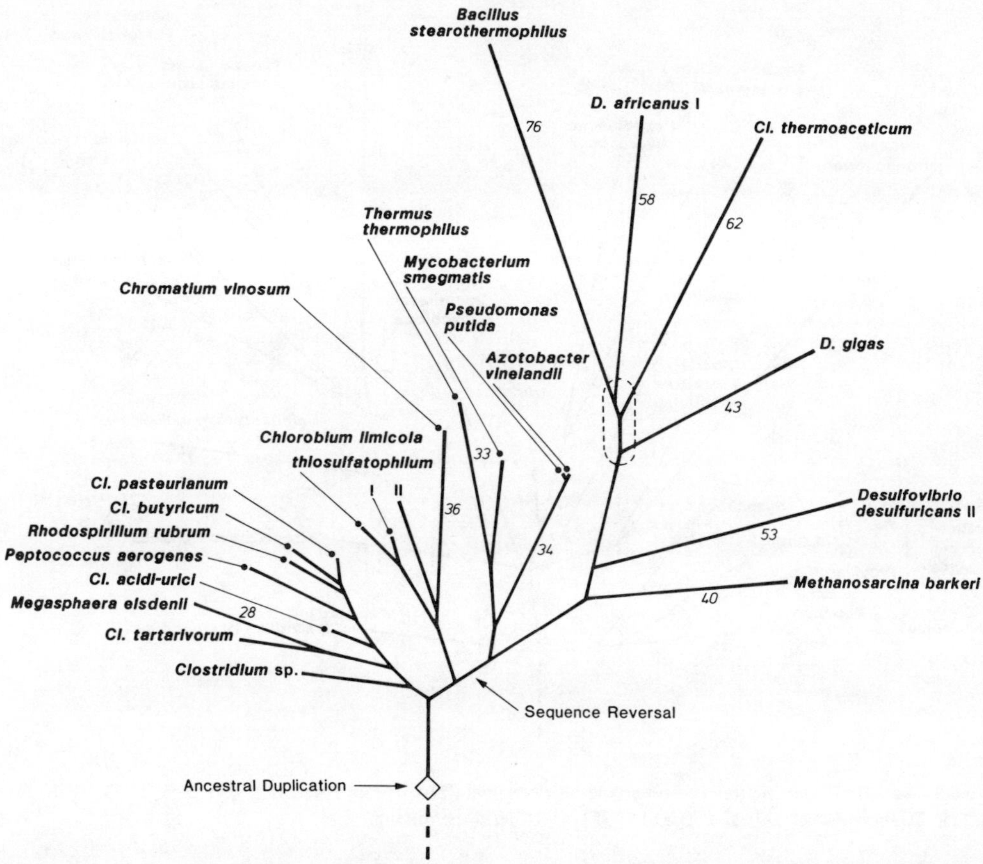

FIGURE 11. Evolutionary tree of bacterial ferredoxins. (From George, D. G., Hunt, L. T., Yeh, L.-S. L., and Barker, W. C., *J. Mol. Evol.*, 22, 20, 1985. With permission.)

discrete exons code for the EGF-like domain[94,95]), and in the vaccinia virus 19-kdalton protein. A single copy of the domain is present in the viral protein and TGF Type 1; the coagulation factors may have one or two copies and EGF has a number of copies, probably arising from internal tandem duplications (the active EGF domain is most similar to the other proteins). Further, in EGF, TGF Type 1, and the 19-kdalton protein a similar potential transmembrane domain follows the EGF-like domain. The evolutionary history of the EGF-like domains is complex, and several duplication events in addition to genetic transfers are involved.[81] The EGF-like domains in all of these proteins probably have a common genetic ancestor.

G. Evolutionary Trees from Viral Proteins

The derivation of evolutionary trees from viral protein sequences may be one of the few practical approaches to studying the evolutionary relationships among viruses and their hosts. Unfortunately the construction and interpretation of such trees present the investigator with a number of problems. Many viral proteins are encoded as very large polyproteins and frequently only much smaller regions of these can be shown to be related; the other regions of these large molecules either had independent evolutionary histories or have changed so rapidly that evidence of their relatedness has been obliterated. Locating sufficiently conserved regions, choosing reasonable boundaries, and constructing an alignment of the distantly related sequences all require considerable judgment on the part of the investigator. Evolu-

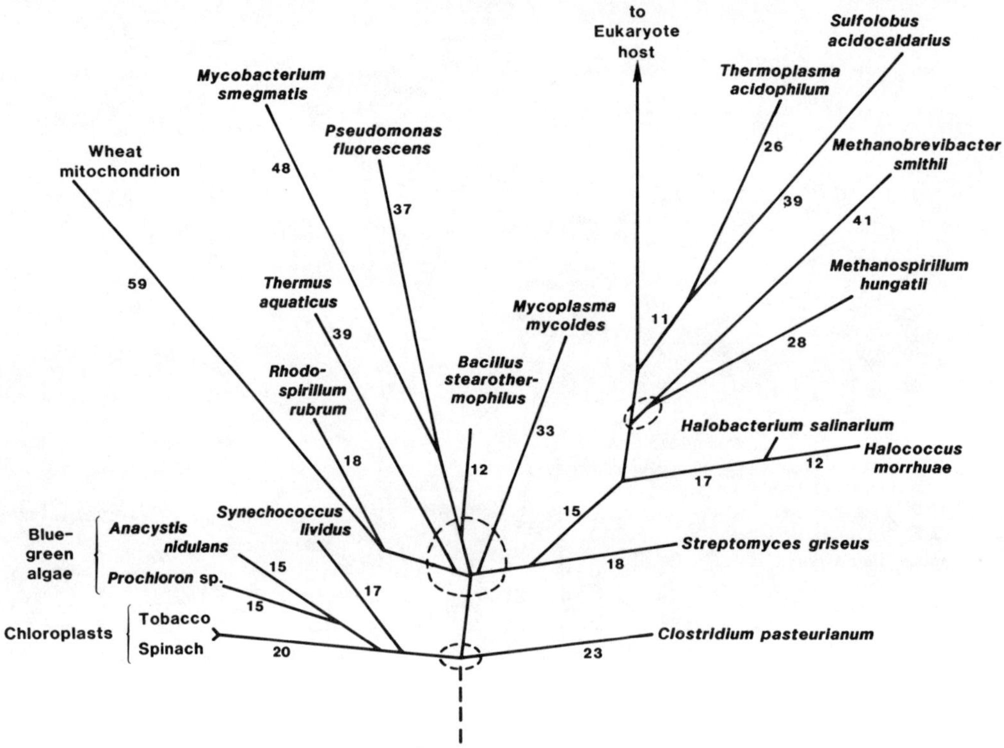

FIGURE 12. Phylogenetic tree of prokaryote 5S ribosomal RNA. (From Hunt, L. T., George, D. G., Yeh, L.-S., and Dayhoff, M. O., *Origins Life,* 14, 657, 1984. With permission.)

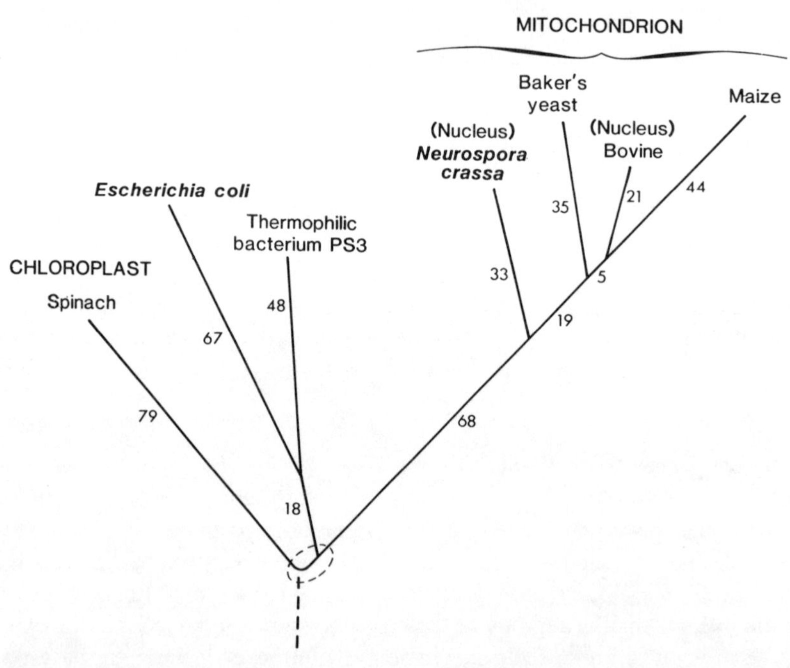

FIGURE 13. Evolutionary tree of ATPase (EC 3.6.1.34) lipid-binding protein. (From Hunt, L. T., George, D. G., and Barker, W. C., *BioSystems,* 18, 223, 1985. With permission.)

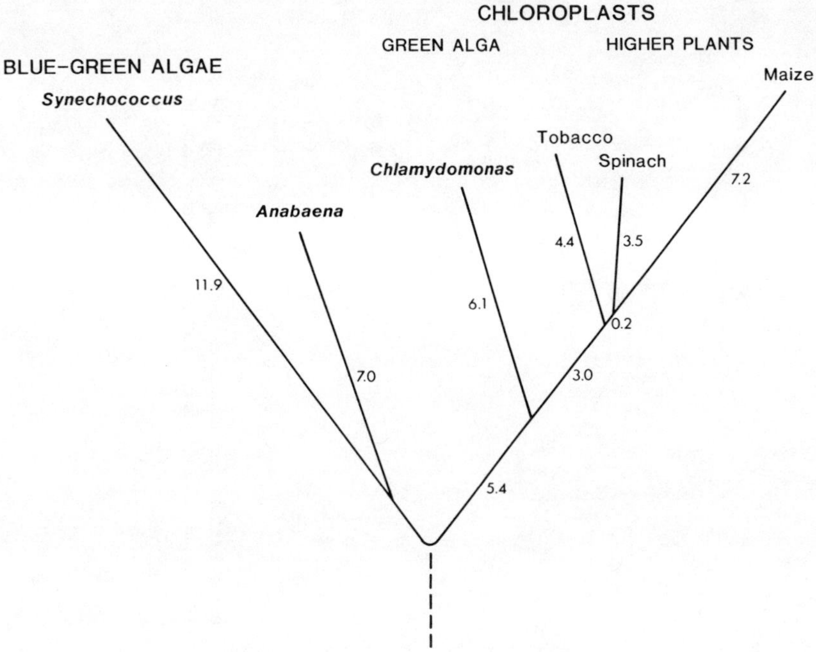

FIGURE 14. Evolutionary tree of ribulose-bisphosphate carboxylase/oxygenase (EC 4.1.1.39), large chain. (From Hunt, L. T., George, D. G., Yeh, L.-S, and Dayhoff, M. O., *Origins Life,* 14, 657, 1984. With permission.)

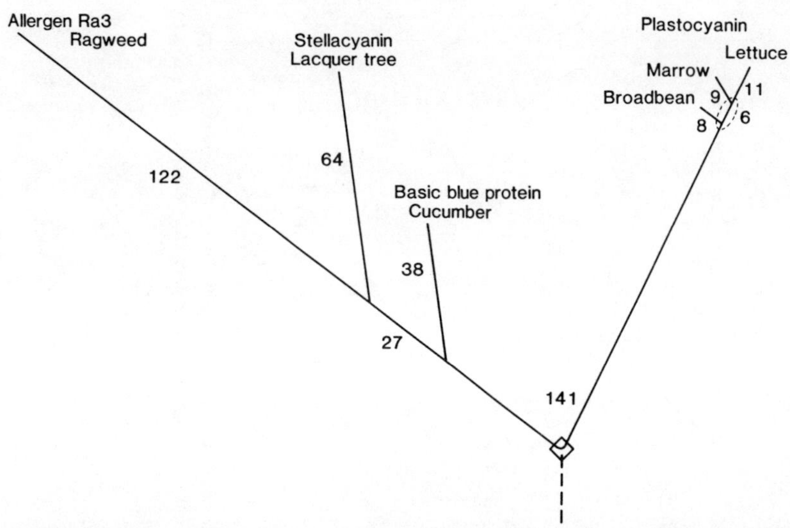

FIGURE 15. Evolutionary tree of some Type 1 copper-binding proteins and ragweed allergen Ra3. (From Hunt, L. T., George, D. G., and Yeh, L.-S. L., *J. Mol. Evol.,* 21, 126, 1985. With permission.)

tionary trees can, of course, also be derived from nucleotide sequences; however, more distant relationships remain detectable in the protein sequences long after they can be seen from the corresponding nucleotide sequences. On the other hand, the branching order of closely related sequences may often be more easily resolved using the nucleotide sequences.

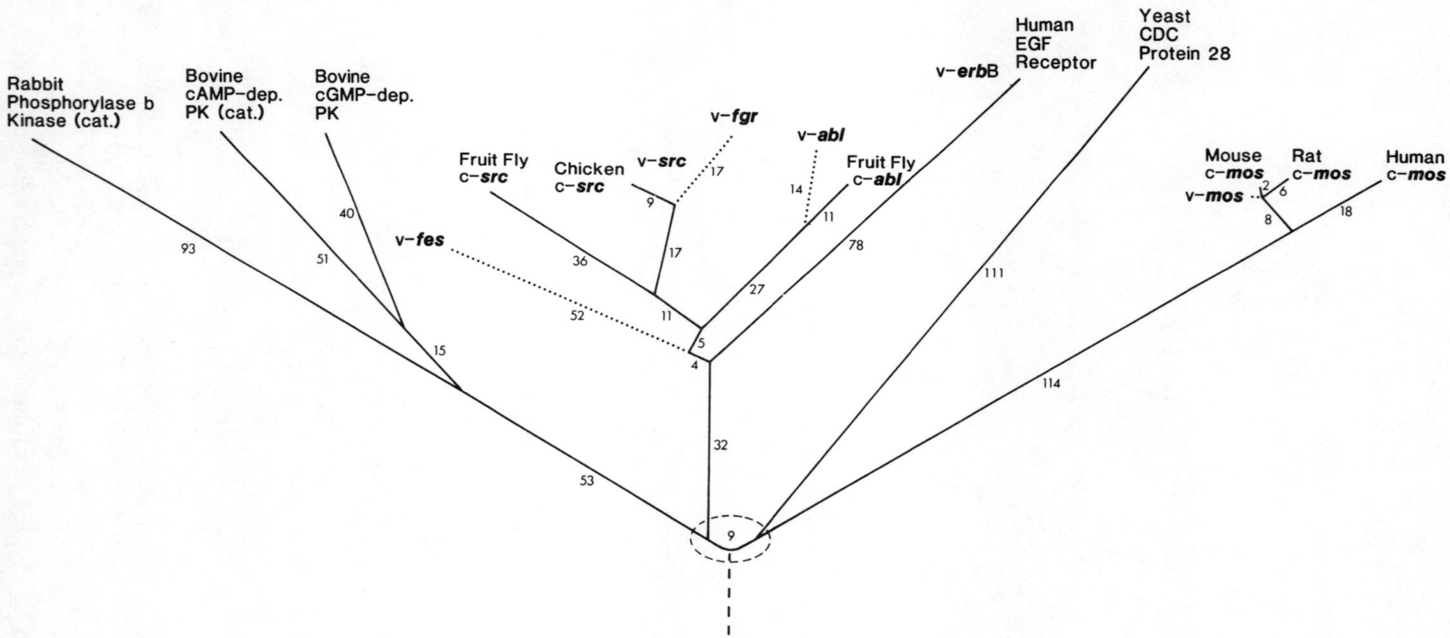

FIGURE 16. Evolutionary tree of the catalytic domains of eukaryote protein kinases and related viral and cellular transforming proteins. Branch lengths are proportional to the estimated amount of evolutionary change expressed as accepted point mutations per 100 residues. The dashed oval indicates some uncertainty in the order of branching.

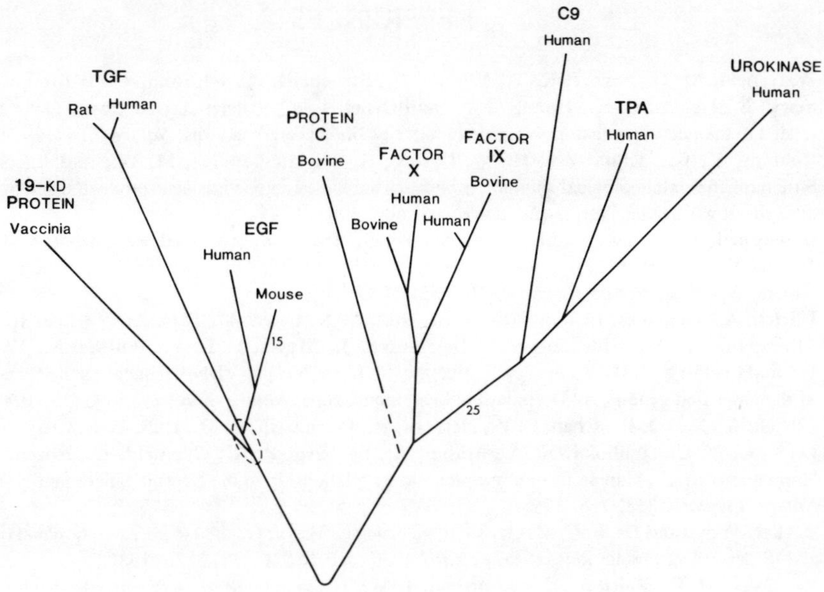

FIGURE 17. Evolutionary tree of type A EGF-like domains.

VII. CONCLUSION

One of the major problems in studying protein evolution is to determine which proteins or regions of proteins are related so that one is able to compare sequences derived from evolutionarily related genetic material. Homologous proteins or domains are those that share a common evolutionary ancestor; they may be encoded by the same gene (or gene segment) in different species or they may be encoded by genes that originated by duplication of an ancestral gene. A newly recognized complication is the extent of transfer of genetic material (within chromosomes, between chromosomes, between organisms) that has been gradually discovered over the past few years. Boundaries of domains are usually arbitrarily chosen and dependent on the kind of domain definition being used; in some cases they appear to correspond to exon-intron junctions. Proteins may show homology only in limited regions or domains rather than along their entire lengths; some multifunctional proteins appear to be mosaics of domains.

Recently, new computer methods have become more widely available for the analysis and comparison of protein sequence data. With the rapid accumulation of these data, the expectation of finding sequences that achieve high similarity scores by chance alone has dramatically increased. As searching algorithms become more efficient and widely used, the finding of unexpected similarities (often mistakenly called ''homologies'') is occurring more frequently. The significance of these similarities is very difficult to assess and statistical measures are often misused and misinterpreted. We have discussed the application of these methods, the evaluation of sequence similarities, and the problem of inferring evolutionary relatedness from similarity measures using a number of examples, including some oncogenes.

ACKNOWLEDGMENTS

We thank Margaret C. Blomquist for editorial suggestions, and Carol E. Toeller for preparation of the figures. This work was supported by NIH grants RR 01821 and HD 09547 and NASA contract NASW-3954.

REFERENCES

1. **Waterfield, M. D., Scrace, G. T., Whittle, N., Stroobant, P., Johnsson, A., Wasteson, A., Westermark, B., Heldin, C.-H., Huang, J. S., and Deuel, T. F.,** Platelet-derived growth factor is structurally related to the putative transforming protein p28^{sis} of simian sarcoma virus, *Nature (London)*, 304, 35, 1983.
2. **Robbins, K. C., Antoniades, H. N., Devare, S. G., Hunkapiller, M. W., and Aaronson, S. A.,** Structural and immunological similarities between simian sarcoma virus gene product(s) and human platelet-derived growth factor, *Nature (London)*, 305, 605, 1983.
3. **Downward, J., Yarden, Y., Mayes, E., Scrace, G., Totty, N., Stockwell, P., Ullrich, A., Schlessinger, J., and Waterfield, M. D.,** Close similarity of epidermal growth factor receptor and v-*erb-B* oncogene protein sequences, *Nature (London)*, 307, 521, 1984.
4. **Ullrich, A., Coussens, L., Hayflick, J. S., Dull, T. J., Gray, A., Tam, A. W., Lee, J., Yarden, Y., Libermann, T. A., Schlessinger, J., Downward, J., Mayes, E. L. V., Whittle, N., Waterfield, M. D., and Seeburg, P. H.,** Human epidermal growth factor receptor cDNA sequence and aberrant expression of the amplified gene in A431 epidermoid carcinoma cells, *Nature (London)*, 309, 418, 1984.
5. **Ullrich, A., Bell, J. R., Chen, E. Y., Herrera, R., Petruzzelli, L. M., Dull, T. J., Gray, A., Coussens, L., Liao, Y.-C., Tsubokawa, M., Mason, A., Seeburg, P. H., Grunfeld, C., Rosen, O. M., and Ramchandran, J.,** Human insulin receptor and its relationship to the tyrosine kinase family of oncogenes, *Nature (London)*, 313, 756, 1985.
6. **Barker, W. C. and Dayhoff, M. O.,** Viral *src* gene products are related to the catalytic chain of mammalian cAMP-dependent protein kinase, *Proc. Natl. Acad. Sci. U.S.A.*, 79, 2836, 1982.
7. **Privalsky, M. L., Ralston, R., and Bishop, J. M.,** The membrane glycoprotein encoded by the retroviral oncogene v-*erb*-B is structurally related to tyrosine-specific protein kinases, *Proc. Natl. Acad. Sci. U.S.A.*, 81, 704, 1984.
8. **Naharro, G., Robbins, K. C., and Reddy, E. P.,** Gene product of v-*fgr onc:* hybrid protein containing a portion of actin and a tyrosine-specific protein kinase, *Science*, 223, 63, 1984.
9. **Heldin, C.-H. and Westermark, B.,** Growth factors: mechanism of action and relation to oncogenes, *Cell*, 37, 9, 1984.
10. **Hunter, T.,** The protein of oncogenes, *Sci. Am.*, 251(2), 70, 1984.
11. **Orcutt, B. C. and Dayhoff, M. O.,** Protein identificaiton system: methods of searching for similar sequences, *IEEE Proc. 7th Annu. Symp. Comput. Appl. Med. Care*, IEEE, Piscataway, N.J., 1983, 579.
12. **Orcutt, B. C. and Barker, W. C.,** Searching the protein sequence database, *Bull. Math. Biol.*, 46, 545, 1984.
13. **Ohta, T.,** Some models of gene conversion for treating the evolution of multigene families, *Genetics*, 106, 517, 1984.
14. **Nagylaki, T.,** The evolution of multigene families under intrachomosomal gene conversion, *Genetics*, 106, 529, 1984.
15. **Nagylaki, T.,** Evolution of multigene families under interchromosomal gene conversion, *Proc. Natl. Acad. Sci. U.S.A.*, 81, 3796, 1984.
16. **Dayhoff, M. O.,** Survey of new data and computer methods of analysis, in *Atlas of Protein Sequence and Structure*, Vol. 5, Suppl. 3, Dayhoff, M. O., Ed., National Biomedical Research Foundation, Washington, D.C., 1979, 1.
17. **Richardson, J. S.,** The anatomy and taxonomy of protein structure, *Adv. Protein Chem.*, 34, 167, 1981.
18. **Kabsch, W. and Sander, C.,** On the use of sequence homologies to predict protein structure: identical pentapeptides can have completely different conformations, *Proc. Natl. Acad. Sci. U.S.A.*, 81, 1075, 1984.
19. **Cantor, C. R.,** The occurrence of gaps in protein sequences, *Biochem. Biophys. Res. Commun.*, 31, 410, 1968.
20. **Haber, J. E. and Koshland, D. E., Jr.,** An evaluation of the relatedness of proteins based on comparison of amino acid sequences, *J. Mol. Biol.*, 50, 617, 1970.
21. **Goad, W. B. and Kanehisa, M. I.,** Pattern recognition in nucleic acid sequences. I. A general method for finding local homologies and symmetries, *Nucleic Acids Res.*, 10, 247, 1982.
22. **Kanehisa, M. I.,** Los Alamos sequence package for nucleic acids and proteins, *Nucleic Acids Res.*, 10, 183, 1982.
23. **Kanehisa, M.,** Ideas: integrated database and extended analysis system for nucleic acids and proteins, *User Man.*, 1984.
24. **McLachlan, A. D.,** Tests for comparing related amino-acid sequences. Cytochrome *c* and cytochrome c551, *J. Mol. Biol.*, 61, 409, 1971.
25. **McLachlan, A. D.,** Repeating sequences and gene duplication in proteins, *J. Mol. Biol.*, 64, 417, 1972.
26. **Dayhoff, M. O., Schwartz, R. M., and Orcutt, B. C.,** A model of evolutionary change in proteins, in *Atlas of Protein Sequence and Structure*, Vol. 5, Suppl. 3, Dayhoff, M. O., Ed., National Biomedical Research Foundation, Washington, D.C., 1979, 345.

27. **Schwartz, R. M. and Dayhoff, M. O.,** Matrices for detecting distant relationships, in *Atlas of Protein Sequence and Structure,* Vol. 5, Suppl. 3, Dayhoff, M. O., Ed., National Biomedical Research Foundation, Washington, D.C., 1979, 353.
28. **Feng, D. F., Johnson, M. S., and Doolittle, R. F.,** Aligning amino acid sequences: comparison of commonly used methods, *J. Mol. Evol.,* 21, 112, 1985.
29. **Barker, W. C., Hunt, L. T., George, D. G., and Orcutt, B. C.,** A resource for protein identification, *CODATA Bull.,* 56, 28, 1984.
30. **Orcutt, B. C., George, D. G., and Dayhoff, M. O.,** Protein and nucleic acid sequence database systems, *Annu. Rev. Biophys. Bioeng.,* 12, 419, 1983.
31. **Dayhoff, M. O., Barker, W. C., and Hunt, L. T.,** Establishing homologies in protein sequences, *Meth. Enzymol.,* 91, 524, 1983.
32. **Lipman, D. J. and Pearson, W. R.,** Rapid and sensitive protein similarity searches, *Science,* 227, 1435, 1985.
33. **Wilbur, W. J. and Lipman, D. J.,** Rapid similarity searches of nucleic acid and protein data banks, *Proc. Natl. Acad. Sci. U.S.A.,* 80, 726, 1983.
34. **Wilbur, W. J. and Lipman, D. J.,** The context dependent comparison of biological sequences, *SIAM J. Appl. Math.,* 44, 557, 1984.
35. **Needleman, S. B. and Wunsch, C. D.,** A general method applicable to the search for similarities in the amino acid sequence of two proteins, *J. Mol. Biol.,* 48, 443, 1970.
36. **Fitch, W. M. and Smith, T. F.,** Optimal sequence alignments, *Proc. Natl. Acad. Sci. U.S.A.,* 80, 1382, 1983.
37. **Fitch, W. M.,** An improved method of testing for evolutionary homology, *J. Mol. Biol.,* 16, 9, 1966.
38. **George, D. G., Yeh, L.-S. L., and Barker, W. C.,** Unexpected relationships between bacteriophage lambda hypothetical proteins and bacteriophage T4 tail-fiber proteins, *Biochem. Biophys. Res. Commun.,* 115, 1061, 1983.
39. **Maizel, J. V., Jr. and Lenk, R. P.,** Enhanced graphic matrix analysis of nucleic acid and protein sequences, *Proc. Natl. Acad. Sci. U.S.A.,* 78, 7665, 1981.
40. **Staden, R.,** An interactive graphics program for comparing and aligning nucleic acid and amino acid sequences, *Nucleic Acids Res.,* 10(9), 2951, 1982.
41. **Hopp, T. P. and Woods, K. R.,** Prediction of protein antigenic determinants from amino acid sequences, *Proc. Natl. Acad. Sci. U.S.A.,* 78, 3824, 1981.
42. **Kyte, J. and Doolittle, R. F.,** A simple method for displaying the hydropathic character of a protein, *J. Mol. Biol.,* 157, 105, 1982.
43. **Chou, P. Y. and Fasman, G. D.,** Empirical predictions of protein conformation, *Annu. Rev. Biochem.,* 47, 251, 1978.
44. **Blomquist, M. C., Hunt, L. T., and Barker, W. C.,** Vaccinia virus 19-kilodalton protein: relationship to several mammalian proteins, including two growth factors, *Proc. Natl. Acad. Sci. U.S.A.,* 81, 7363, 1984.
45. **Brown, J. P., Twardzik, D. R., Marquardt, H., and Todaro, G. J.,** Vaccinia virus encodes a polypeptide homologous to epidermal growth factor and transforming growth factor, *Nature (London),* 313, 491, 1985.
46. **Reisner, A. H.,** Similarity between the vaccinia virus 19K early protein and epidermal growth factor, *Nature (London),* 313, 801, 1985.
47. **Marquardt, H., Hunkapiller, M. W., Hood, L. E., Twardzik, D. R., De Larco, J. E., Stephenson, J. R., and Todaro, G. J.,** Transforming growth factors produced by retrovirus-transformed rodent fibroblasts and human melanoma cells: amino acid sequence homology with epidermal growth factor, *Proc. Natl. Acad. Sci. U.S.A.,* 80, 4684, 1983.
48. **Marquardt, H., Hunkapiller, M. W., Hood, L. E., and Todaro, G. J.,** Rat transforming growth factor type 1: structure and relation to epidermal growth factor, *Science,* 223, 1079, 1984.
49. **Devare, S. G., Reddy, E. P., Law, J. D., Robbins, K. C., and Aaronson, S. A.,** Nucleotide sequences of the simian sarcoma virus genome: demonstration that its acquired cellular sequences encode the transforming gene product $p28^{sis}$, *Proc. Natl. Acad. Sci. U.S.A.,* 80, 731, 1983.
50. **Antoniades, H. N. and Hunkapiller, M. W.,** Human platelet-derived growth factor (PDGF): amino-terminal amino acid sequence, *Science,* 220, 963, 1983.
51. **Doolittle, R. F., Hunkapiller, M. W., Hood, L. E., Devare, S. G., Robbins, K. C., Aaronson, S. A., and Antoniades, H. N.,** Simian sarcoma virus *onc* gene, v-*sis*, derived from the gene (or genes) encoding a platelet-derived growth factor, *Science,* 221, 275, 1983.
52. **Josephs, S. F., Ratner, L., Clarke, M. F., Westin, E. H., Reitz, M. S., and Wong-Staal, F.,** Transforming potential of human c-*sis* nucleotide sequences encoding platelet-derived growth factor, *Science,* 225, 636, 1984.
53. **Yamamoto, T., Nishida, T., Miyajima, N., Kawai, S., Ooi, T., and Toyoshima, K.,** The *erbB* gene of avian erythroblastosis virus is a member of the *src* gene family, *Cell,* 35, 71, 1983.

54. **Gilmore, T., DeClue, J. E., and Martin, G. S.,** Protein phosphorylation at tyrosine is induced by the v-*erbB* gene product in vivo and in vitro, *Cell,* 40, 609, 1985.
55. **Kris, R. M., Lax, I., Gullick, W., Waterfield, M. D., Ullrich, A., Fridkin, M., and Schlessinger, J.,** Antibodies against a synthetic peptide as a probe for the kinase activity of the avian EGF receptor and v-erbB protein, *Cell,* 40, 619, 1985.
56. **Kitamura, N., Kitamura, A., Toyoshima, K., Hirayama, Y., and Yoshida, M.,** Avian sarcoma virus Y73 genome sequence and structural similarity of its transforming gene product to that of Rous sarcoma virus, *Nature (London),* 297, 205, 1982.
57. **Schlesinger, D. H. and Goldstein, G.,** Molecular conservation of 74 amino acid sequence of ubiquitin between cattle and man, *Nature (London),* 255, 423, 1975.
58. **Gavilanes, J. G., Gonzalez de Buitrago, G., Perez-Castells, R., and Rodriquez, R.,** Isolation, characterization, and amino acid sequence of a ubiquitin-like protein from insect eggs, *J. Biol. Chem.,* 257, 10267, 1982.
59. **Hunt, L. T. and Dayhoff, M. O.,** Evolution of chromosomal proteins, in *Macromolecular Sequences in Systematic and Evolutionary Biology,* Goodman, M., Ed., Plenum Press, New York, 1982, 193.
60. **Olson, M. O. J., Goldknopf, I. L., Guetzow, K. A., James, G. T., Hawkins, T. C., Mays-Rothberg, C. J., and Busch, H.,** The NH_2 and COOH-terminal amino acid sequence of nuclear protein A24, *J. Biol. Chem.,* 251, 5901, 1976.
61. **Hunt, L. T. and Dayhoff, M. O.,** Amino-terminal sequence identity of ubiquitin and the nonhistone component of nuclear protein A24, *Biochem. Biophys. Res. Commun.,* 74, 650, 1977.
62. **Goldknopf, I. L. and Busch, H.,** Isopeptide linkage between nonhistone and histone 2A polypeptides of chromosomal conjugate-protein A24, *Proc. Natl. Acad. Sci. U.S.A.,* 74, 864, 1977.
63. **Seidah, N. G., Crine, P., Benjannet, S., Scherrer, H., and Chretien, M.,** Isolation and partial characterization of a biosynthetic N-terminal methionyl peptide of bovine pars intermedia: relationship to ubiquitin, *Biochem. Biophys. Res. Commun.,* 80, 600, 1978.
64. **Dayhoff, M. O. and Orcutt, B. C.,** Methods for identifying proteins by using partial sequences, *Proc. Natl. Acad. Sci. U.S.A.,* 76, 2170, 1979.
65. **Grubb, A. and Lofberg, H.,** Human gamma-trace, a basic microprotein: amino acid sequence and presence in the adenohypophysis, *Proc. Natl. Acad. Sci. U.S.A.,* 79, 3024, 1982.
66. **Isemura, S., Saitoh, E., and Sanada, K.,** Isolation and amino acid sequence of SAP-1, an acidic protein of human whole saliva, and sequence homology with human gamma-trace, *J. Biochem.,* 96, 489, 1984.
67. **Schwabe, C., Anastasi, A., Crow, H., McDonald, J. K., and Barrett, A. J.,** Cystatin: amino acid sequence and possible secondary structure, *Biochem. J.,* 217, 813, 1984.
68. **Turk, V., Brzin, J., Longer, M., Ritonja, A., and Eropkin, M.,** Protein inhibitors of cysteine proteinases. III. Amino-acid sequence of cystatin from chicken egg white, *Hoppe-Seyler's Z. Physiol. Chem.,* 364, 1487, 1983.
69. **Brzin, J., Popovic, T., and Turk, V.,** Human cystatin, a new protein inhibitor of cysteine proteinases, *Biochem. Biophys. Res. Commun.,* 118, 103, 1984.
70. **Isemura, S., Saitoh, E., Ito, S, Isemura, M., and Sanada, K.,** Cystatin S: a cysteine proteinase inhibitor of human saliva, *J. Biochem.,* 96, 1311, 1984.
71. **Takio, K., Wade, R. D., Smith, S. B., Krebs, E. G., Walsh, K. A., and Titani, K.,** Guanosine cyclic 3′,5′-phosphate dependent protein kinase, a chimeric protein homologous with two separate protein families, *Biochemistry,* 23, 4207, 1984.
72. **Reimann, E. M., Titani, K., Ericsson, L. H., Wade, R. D., Fischer, E. H., and Walsh, K. A.,** Homology of the gamma subunit of phosphorylase *b* kinase with cAMP-dependent protein kinase, *Biochemistry,* 23, 4185, 1984.
73. **Lorincz, A. T. and Reed, S. I.,** Primary structure homology between the product of yeast cell division control gene *CDC28* and vertebrate oncogenes, *Nature (London),* 307, 183, 1984.
74. **Titani, K., Sasagawa, T., Ericsson, L. H., Kumar, S., Smith, S. B., Krebs, E. G., and Walsh, K. A.,** Amino acid sequence of the regulatory subunit of bovine type I adenosine cyclic 3′,5′-phosphate dependent protein kinase, *Biochemistry,* 23, 4193, 1984.
75. **Weber, I. T., Takio, K., Titani, K., and Steitz, T. A.,** The cAMP-binding domains of the regulatory subunit of cAMP-dependent protein kinase and the catabolite gene activator protein are homologous, *Proc. Natl. Acad. Sci. U.S.A.,* 79, 7679, 1982.
76. **Yamamoto, T., Davis, C. G., Brown, M. S., Schneider, W. J., Casey, M. L., Goldstein, J. L., and Russell, D. W.,** The human LDL receptor: a cysteine-rich protein with multiple Alu sequences in its mRNA, *Cell,* 39, 27, 1984.
77. **Sudhof, T. C., Goldstein, J. L., Brown, M. S., and Russell, D. W.,** The LDL receptor gene: a mosaic of exons shared with different proteins, *Science,* 228, 815, 1985.
78. **Stanley, K. K., Kocher, H.-P., Luzio, J. P., Jackson, P., and Tschopp, J.,** The sequence and topology of human complement component C9, *EMBO J.,* 4(2), 375, 1985.

79. **Rosenfeld, S. I., Packman, C. H., and Leddy, J. P.,** Inhibition of the lytic action of cell-bound terminal complement components by human high density lipoproteins and apoproteins, *J. Clin. Invest.*, 71, 795, 1983.
80. **Sudhof, T. C., Russell, D. W., Goldstein, J. L., Brown, M. S., Sanchez-Pescador, R., and Bell, G. I.,** Cassette of eight exons shared by genes for LDL receptor and EGF precursor, *Science*, 228, 893, 1985.
81. **Doolittle, R. F., Feng, D. F., and Johnson, M. S.,** Computer-based characterization of epidermal growth factor precursor, *Nature (London)*, 307, 558, 1984.
82. **Delbaere, L. T. J., Hutcheon, W. L. B., James, M. N. G., and Thiessen, W. E.,** Tertiary structural differences between microbial serine proteases and pancreatic serine enzymes, *Nature (London)*, 257, 758, 1975.
83. **Brayer, G. D., Delbaere, L. T. J., and James, M. N. G.,** Molecular structure of crystalline *Streptomyces griseus* protease A at 2.8 A resolution. II. Molecular conformation, comparison with alpha-chymotrypsin and active-site geometry, *J. Mol. Biol.*, 124, 261, 1978.
84. **Neurath, H.,** Evolution of proteolytic enzymes, *Science*, 224, 350, 1984.
85. **Magnusson, S., Sottrup-Jensen, L., Petersen, T. E., and Claeys, H.,** The primary structure of prothrombin, the role of vitamin K in blood coagulation and a thrombin catalyzed 'negative feed-back' control mechanism for limiting the activation of prothrombin, in *Boerhaave Symposium on Prothrombin and Related Coagulation Factors*, Hemiker, H. C. and Veltkamp, J. J., Eds., Leiden University Press, Leiden, The Netherlands, 1975, 25.
86. **Walz, D. A., Hewett-Emmett, D., and Seegers, W. H.,** Amino acid sequence of human prothrombin fragments 1 and 2, *Proc. Natl. Acad. Sci. U.S.A.*, 74, 1969, 1977.
87. **Fernlund, P. and Stenflo, J.,** Amino acid sequence of the light chain of bovine protein C, *J. Biol. Chem.*, 257, 12170, 1982.
88. **Katayama, K., Ericsson, L. H., Enfield, D. L., Walsh, K. A., Neurath, H., Davie, E. W., and Titani, K.,** Comparison of amino acid sequence of bovine coagulation factor IX (Christmas factor) with that of other vitamin K-dependent plasma proteins, *Proc. Natl. Acad. Sci. U.S.A.*, 76, 4990, 1979.
89. **Enfield, D. L., Ericsson, L. H., Fujikawa, K., Walsh, K. A., Neurath, H., and Titani, K.,** Amino acid sequence of the light chain of bovine factor X_1 (Stuart factor), *Biochemistry*, 19, 659, 1980.
90. **McMullen, B. A., Fujikawa, K., Kisiel, W., Sasagawa, T., Howald, W. N., Kwa, E. Y., and Weinstein, B.,** Complete amino acid sequence of the light chain of human blood coagulation factor X: evidence for identification of residue 63 as beta-hydroxyaspartic acid, *Biochemistry*, 22, 2875, 1983.
91. **Young, C. L., Barker, W. C., Tomaselli, C. M., and Dayhoff, M. O.,** Serine Proteases, in *Atlas of Protein Sequence and Structure*, Vol. 5, Suppl. 3, Dayhoff, M. O., Ed., National Biomedical Research Foundation, Washington, D.C., 1979, 73.
92. **Gunzler, W. A., Steffens, G. J., Otting, F., Kim, S.-M. A., Frankus, E., and Flohe, L.,** The primary structure of high molecular mass urokinase from human urine. The complete amino acid sequence of the A chain, *Hoppe-Seyler's Z. Physiol. Chem.*, 363, 1155, 1982.
93. **Pennica, D., Holmes, W. E., Kohr, W. J., Harkins, R. N., Vehar, G. A., Ward, C. A., Bennett, W. F., Yelverton, E., Seeburg, P. H., Heyneker, H. L., Goeddel, D. V., and Collen, D.,** Cloning and expression of human tissue-type plasminogen activator cDNA in *E. coli*, *Nature (London)*, 301, 214, 1983.
94. **Ny, T., Elgh, F., and Lund, B.,** The structure of the human tissue-type plasminogen activator gene: correlation of intron and exon structures to functional and structural domains, *Proc. Natl. Acad. Sci. U.S.A.*, 81, 5355, 1984.
95. **Anson, D. S., Choo, K. H., Rees, D. J. G., Giannelli, F., Gould, K., Huddleston, J. A., and Brownlee, G. G.,** The gene structure of human anti-haemophilic factor IX, *EMBO J.*, 3, 1053, 1984.
95a. **Johnson, G. C. and Barker, W. C.,** unpublished results.
96. **Sottrup-Jensen, L., Claeys, H., Zajdel, M., Petersen, T. E., and Magnusson, S.,** The primary structure of human plasminogen: isolation of two lysine-binding fragments and one "mini-" plasminogen (MW, 38,000) by elastase-catalyzed-specific limited proteolysis, in *Progress in Chemical Fibrinolysis and Thrombolysis*, Vol. 3, Davidson, J. F., Rowan, R. M., Samama, M. M., and Desnoyers, P. C., Eds., Raven Press, New York, 1978, 191.
97. **Degen, S. J. F., MacGillivray, R. T. A., and Davie, E. W.,** Characterization of the complementary deoxyribonucleic acid and gene coding for human prothrombin, *Biochemistry*, 22, 2087, 1983.
98. **Banyai, L., Varadi, A., and Patthy, L.,** Common evolutionary origin of the fibrin-binding structures of fibronectin and tissue-type plasminogen activator, *FEBS Lett.*, 163, 37, 1983.
99. **Petersen, T. E., Thogersen, H. C., Skorstengaard, K., Vibe-Pedersen, K., Sahl, P., Sottrup-Jensen, L., and Magnusson, S.,** Partial primary structure of bovine plasma fibronectin: three types of internal homology, *Proc. Natl. Acad. Sci. U.S.A.*, 80, 137, 1983.
100. **Garcia-Pardo, A., Pearlstein, E., and Frangione, B.,** Primary structure of human plasma fibronectin. The 29,000-dalton NH_2-terminal domain, *J. Biol. Chem.*, 258, 12670, 1983.

101. **Lozier, J., Takahashi, N., and Putnam, F. W.,** Complete amino acid sequence of human plasma beta$_2$-glycoprotein I, *Proc. Natl. Acad. Sci. U.S.A.,* 81, 3640, 1984.
102. **Morley, B. J. and Campbell, R. D.,** Internal homologies of the Ba fragment from human complement component factor B, a class III MHC antigen, *EMBO J.,* 3, 153, 1984.
103. **Mole, J. E., Anderson, J. K., Davison, E. A., and Woods, D. E.,** Complete primary structure for the zymogen of human complement factor B, *J. Biol. Chem.,* 259, 3407, 1984.
104. **Kristensen, T., Wetsel, R. A., and Tack, B. F.,** Primary structure of the human complement control protein factor H, *Fed. Proc.,* 44, 1531, 1985.
105. **Reid, K. B. M.,** Complete amino acid sequence of human C4 binding protein, F.A.S.E.B. 69th Annu. Meet., Anaheim, Calif., April 1985.
106. **Gilbert, W.,** Genes-in-pieces revisited, *Science,* 228, 823, 1985.
107. **Dayhoff, M. O.,** The origin and evolution of protein superfamilies, *Fed. Proc.,* 35, 2132, 1976.
108. **Schwartz, R. M. and Dayhoff, M. O.,** Origins of prokaryotes, eukaryotes, mitochondria, and chloroplasts, *Science,* 199, 395, 1978.
109. **Fitch, W. M. and Margoliash, E.,** Construction of phylogenetic trees, *Science,* 155, 279, 1967.
110. **Dayhoff, M. O., Park, C. M., and McLaughlin, P. J.,** Building a phylogenetic tree: cytochrome c, in *Atlas of Protein Sequence and Structure,* Vol. 5, Dayhoff, M. O., Ed., National Biomedical Research Foundation, Washington, D.C., 1972, 7.
111. **Peacock, D. and Boulter, D.,** Use of amino acid sequence data in phylogeny and evaluation of methods using computer simulation, *J. Mol. Biol.,* 95, 513, 1975.
112. **Astolfi, P., Kidd, K. K., and Cavalli-Sforza, L. L.,** A comparison of methods for reconstructing evolutionary trees, *Syst. Zool.,* 30, 156, 1981.
113. **Blanken, R. L., Klotz, L. C., and Hinnebusch, A. G.,** Computer comparison of new and existing criteria for constructing evolutionary trees from sequence data, *J. Mol. Evol.,* 19, 9, 1982.
114. **Barker, W. C. and Dayhoff, M. O.,** Identifying unknown proteins, *IEEE 1983 Proc. 7th Annu. Symp. Comput. Appl. Med. Care,* IEEE, Piscataway, N.J., 1983, 584.
115. **Moore, D. D., Conkling, M. A., and Goodman, H. M.,** Human growth hormone: a multigene family, *Cell,* 29, 285, 1982.
116. **Barnabas, J., Schwartz, R. M., and Dayhoff, M. O.,** Evolution of major metabolic innovations in the precambrian, *Origins Life,* 12, 81, 1982.
117. **Dayhoff, M. O.,** Evolutionary connections of biological kingdoms based on protein and nucleic acid sequence evidence, *Precamb. Res.,* 20, 299, 1983.
118. **Schwartz, R. M. and Dayhoff, M. O.,** Chloroplast origins: inferences from protein and nucleic acid sequences, *Ann. N.Y. Acad. Sci.,* 361, 260, 1981.
119. **George, D. G., Hunt, L. T., Yeh, L.-S. L., and Barker, W. C.,** New perspectives on bacterial ferredoxin evolution, *J. Mol. Evol.,* 22, 20, 1985.
120. **Hunt, L. T., George, D. G., Yeh, L.-S., and Dayhoff, M. O.,** Evolution of prokaryote and eukaryote lines inferred from sequence evidence, *Origins Life,* 14, 657, 1984.
121. **Hunt, L. T., George, D. G., and Yeh, L.-S. L.,** Ragweed allergen Ra3: relationship to some type 1 copper-binding proteins, *J. Mol. Evol.,* 21, 126, 1985.
122. **Massague, J.,** Epidermal growth factor-like transforming growth factor. I. Isolation, chemical characterization, and potentiation by other transforming factors from feline sarcoma virus-transformed rat cells, *J. Biol. Chem.,* 258, 13606, 1983.

Chapter 2

PROTEIN STRUCTURE AND ONCOGENESIS: A STUDY OF THE P21 PROTEIN

Matthew R. Pincus

TABLE OF CONTENTS

I. INTRODUCTION

A crucial recent discovery that sheds much light on oncogenesis in humans was that a single gene, called the oncogene, cloned from the DNA of a human bladder carcinoma (EJ gene) caused transformation of cells in culture (NIH 3T3 cells) into which it was transfected.[1,2] This gene is a member of the *ras* (rat sarcoma) — gene family, originally identified in oncogenic retroviruses. Hybridization of this gene with normal human DNA indicates that it binds specifically to one segment of the normal human DNA and that it was therefore virtually identical to a single normal human gene,[1,2] called the proto-oncogene. Restriction endonuclease digestion of the oncogene and proto-oncogene followed by cross ligations (splicing) of DNA segments of proto-oncogene and oncogene to create molecular hybrids and transfection of these spliced hybrid genes into NIH 3T3 cells indicated that, in order for the gene to transform the cells, an abnormal portion containing the 12th coding triplet in the first exon of the abnormal gene was required.[1,2] Detailed sequencing of both normal and abnormal genes revealed that the only difference between the two genes indeed occurred at the 12th coding triplet in which GGC in the proto-oncogene was converted into GTC in the onogene.[1-3] In the protein product, a Gly residue is converted to Val at position 12 in the sequence consisting of 189 amino acid residues. The protein coded for by each gene has a molecular weight of 21,000 daltons, and is called, therefore, the P21 protein. The level of expression of both genes is virtually the same, i.e., the amount of P21 protein product formed as a function of time is the same in cells containing either gene. Thus, the transforming agent appears to be the oncogene-encoded protein product rather than the gene itself.[1,2] This conclusion has received firm support from microinjection experiments.[4] Microinjection of purified oncogene-encoded protein (with Val at position 12) obtained from overexpression of the oncogene in *Escherichia coli* into NIH 3T3 cells results in transient cell transformation.[4] However, after the protein is degraded intracellularly, the cells revert back to a normal phenotype. Microinjection of the proto-oncogene product causes no cell transformation.[4] Thus, a single amino acid substitution (Gly → Val) at position 12 in the P21 protein can cause normal cells to become malignant. It is of great interest to determine, therefore, the effects such a substitution can have on protein structure and function.

The methods of conformational analysis have been applied to determination of the effects of transforming substitutions on the structure of the relevant peptides in which they occur (*vide infra*). The results indicate that substitution of all L-amino acids except for Pro and Gly 12 all result in a well-defined structural change in the transforming segment. Only Gly and Pro at position 12 are capable of promoting a unique structure associated with normal protein function. These predictions as to substitutions that can cause malignant transformation of cells have been confirmed experimentally. Before discussing the specific results using this approach, the properties of the P21 protein are discussed, followed by a discussion of the methods used to compute the three-dimensional structures of peptides from their amino acid sequences.

A. Characteristics of the P21 Proteins

The P21 proteins are now known to be contained in all eukaryotic cells and are membrane-bound to the inner membrane surface. Unlike for hemoglobin where a single specific Val-for -Glu subsitution occurs at a position 6 resulting in sickle cell anemia (other substitutions do not result in abnormal hemoglobins), it is known that arbitrary amino acid substitutions for Gly 12 in the P21 protein result in a transforming protein. Thus, for example, substitution of Arg (Harvey),[1,2] Lys,[1,2] Ser[1,2] (Kirsten),[1,2] and Asp[5] for Gly 12 always results in a transforming product. This result suggests that substitutions for Gly 12 result in unique structural changes in the protein. The substitutions at position 12 occur in the middle of a hydrophobic decapeptide Leu 6-Gly 15, the transforming region. The sequence for the first 30 amino acid residues of the P21 protein is[1-3]

```
Met-Thr-Glu-Tyr-Lys-Leu-Val-Val-Val-Gly-        10
Ala-Gly-Gly-Val-Gly-Lys-Ser-Ala-Leu-Thr-        20
Ile-Glu-Leu-Ile-Gln-Asn-His-Phe-Val-Asp-        30
```

The underlined Gly is the critical residue at position 12. As discussed at greater length below substitutions for Gly 13 can also be transforming.[6]

It is now known that the P21 proteins bind with high specificity and affinity to GTP.[7,8] All proteins regardless of the substitutions at position 12 bind to GTP with high affinity.[7,8] However, the normal protein contains GTPase activity which is approximately tenfold greater than that for the substituted proteins.[8]

It is hypothesized that the P21 proteins act as G-proteins, i.e., they bind to GTP and interact with an adenyl cyclase system which results in synthesis of cyclic AMP (cAMP).[9] The cAMP acts as a "second messenger" that activates the phosphorylation of critical, probably membrane, proteins, which in an as yet unknown manner triggers cell division.[9] That the P21 proteins are cirtical in regulation of cell division has been demonstrated in yeast cells, the simplest eukaryotic cells. Mutant cells from which the gene coding for yeast P21 proteins has been deleted are incapable of dividing completely.[10]

It has been further hypothesized that P21 proteins, with substitutions for Gly 12, may perpetually stimulate the cyclase systems and thus continually stimulate cell division because they hydrolyze GTP to GDP and Pi only slowly as compared with the normal P21 protein. While this hypothesis is an attractive and simple one, it cannot be completely correct because, as will be discussed below, high intracellular levels of the *normal* protein also cause cell transformation.[11] This latter condition would be expected to give rise to *higher* (not lower) GTPase levels than normal.

B. Domain Structure of the P21 Proteins

1. Amino Terminal Segment Containing the Transforming Region

A number of different studies suggest that the P21 proteins contain long polypeptide segments that are structurally and functionally independent. It has been demonstrated that the isolated peptide segment 4-21 synthesized using solid phase methods can bind GTP with a high affinity although about 100 times lower than that for P21 protein.[11a] Thus, this segment is strongly implicated in GTP binding. This conclusion was reached on the basis of a number of studies on proteins with sequences that are homologous to the transforming region.[12,13] Each of these proteins is known to bind nucleotides (ATP, GTP, or NAD), and the homologous segments in proteins of known three-dimensional structure are known to bind to the nucleotides with which the protein interacts.

On the basis of the homology of the 4-27 sequence of the P21 protein (proto-oncogene-encoded) to a similar sequence in a nucleotide binding protein of known structure, *p*-hydroxybenzoate hydroxylase (PHBH), a model was propsed for how subsititions of L-amino acids for Gly 12 affect protein activity.[13] Construction of a model for P21 protein based on the nucleotide binding region of PHBH seemed to show that any side chain of an amino acid would interfere with phosphate binding and hence would result in lowered affinities for nucleotide binding. In the absence of nucleotide binding, the P21 protein, postulated to be a regulatory subunit of an enzyme, would lose its ability to inhibit enzyme activity.[13] As discussed above, it is now known that P21 proteins with substitutions at position 12 all bind GTP with high affinity.[7,8] Furthermore, as discussed in detail below, substitution of Pro for Gly 12 results in a *nontransforming* protein.[14] Finally, the sequence of PHBH that is homologous to the 4-27 sequence of the P21 protein is also homologous to other sequences of proteins whose structures are known such as lactate dehydrogenase (LDH).[13] However, the specific structures of the homologous segments for the two proteins PHBH and LDH are quite different, and it is difficult to determine which if either of the two structures is the more "correct" one. Thus, model-building studies may lead to erroneous results. However, the discovery of these important homologies clearly implicates the 4-27 segment of the P21 protein as being involved in nucleotide binding.

2. Sequence Around Position 61

In addition to transforming substitutions in the amino terminal transforming hydrophobic decapeptide, a number of other amino acid substitutions are known that result in activation of the P21 protein, specifically at position 59, 61, and 63. Substitution of Leu or Lys for Gln at position 61 results in a strongly transforming protein.[6,15] The substitution of Thr for Ala at position 59 also results in cell transformation.[6] In this case, it is known that Thr 59 becomes phosphorylated by GTP in an autophosphorylation reaction, possibly implicating the latter process as being responsible for the transforming properties of the protein.[6] These findings strongly implicate residue 59 and surrounding residues (such as 61) as being intimately involved in phosphate binding.

Substitution of Lys for Glu 63 in the P21 protein also results in cell transformation.[6] Double mutagenesis experiments have been performed resulting in two critical substitutions in the P21 proteins at positions 12 (Val for Gly) and 63 (Lys for Glu).[6] This protein exhibits a significantly lower transforming potential than either protein with a single substitution at either position 12 or 63.[6] This result suggests that the peptide segment involving residues 59-63 may lie near the amino terminal transforming region so that both peptide segments may be involved in GTP binding.

Further searches for homology betewen the P21 protein and other proteins have revealed a weak similarity in sequences between the residues from 119-133 in P21 and a segment of a G protein involved in protein synthesis, elongation factor tu (EF-tu).[16] The corresponding segment of EF-tu is known to form part of a GTP nucleotide binding site so that a similar function may be attributable to the 199-133 sequence in P21 protein. This segment may further lie near the other two segments known to be involved in nucleotide binding described above. However, the Ef-tu molecule contains no sequence homology to the latter two regions of P21 and therefore no firm conclusions about the structure and interactions of the 119-133 sequence in P21 can be drawn.

3. The Carboxyl Terminal Region

As mentioned above, the P21 proteins are known to bind to the inner cell membrane in cells. A number of studies have now indicated that the binding takes place via the carboxy terminal end.[17,18] In all the P21 proteins there is an invariant Cys residue three residues from the carboxyl terminus. In mammalian varieties of P21 which contain 189 (or 188) residues this residue is Cys 186. (In the yeast P21 protein approximately 100 amino acid residues are inserted after residue 160.[10]) Genetic recombination experiments have been performed in which Cys 186 has been replaced with a Ser residue.[17] The resulting protein is not capable of binding to the inner cell membrane and exists completely within the cell cytosol.[17,18] While these proteins, which contain transforming substitutions, bind with high affinity to GTP, they do not cause cell transformation,[17,18] a finding that establishes membrane binding as necessary for cell transformation. It has recently been extablished that Cys 186 is necessary for membrane binding because palmitate is added to it (as a thiol ester) in a post-translational modification of the protein,[17,18] as has been observed in other animal systems.[19]

It has further been shown that up to 14 amino acid residues between residues 165 and 180 can be deleted with no loss of the ability of the protein to bind to the membrane or to transform cells.[17,18] This finding sugests that the membrane-binding portion of the molecule is not closely associated with the rest of the protein and constitutes a separate domain. However, substitutions and deletions in the carboxyl terminal decapeptide 180-189 do result in diminished or absent membrane binding.[17,18,20] This result is of great interest in view of the fact, as noted above, that this region of the P21 protein is highly variable in sequence among species with Cys 186 as the only invariant residue. The carboxyl terminal decapeptide of four different P21 proteins encoded by Harvey (H)-, Kirsten (KA and KB)-, and N-*ras* genes are listed below.

Gly-Cys-Met-Ser-Cys-Lys-Cys-Val-Leu-Ser	H
Cys-Val-Lys-Ile-Lys-Lys-Cys-Ile-Ile-Met	KA
Lys-Lys-Ser-Lys-Thr-Lys-Cys-Val-Ile-Met	KB
Lys-Lys-Ser-Gly-Leu-Pro-Cys-Val-Val-Met	N

Experiments on the H sequence[17,18,20] suggest that these decapeptides contain structural requirements without which membrane binding does not occur. In the process of deleting gene sequences coding for the amino acids between residues 165 and 180, the codons for Pro-Asp-Glu become inserted between these residues.[17,18,20] It has been found that a protein (PBW766 mutant of H-*ras*) with deletion of residues 165-183 and replacement with Pro-Asp-Glu binds only partially to the inner cell membrane but still retains its transforming ability. Replacement of residues 181-183 by Pro-Asp-Glu (PBW768 mutant) results in loss of membrane binding and all transforming activity while substitution of this sequence for residues 177-179 (PBW757 mutant) causes no loss of membrane binding or transforming activity. Also, substitution of Thr-Pro for residues 187-189 (PBW277 mutant) results in loss of all membrane binding and transforming activity.[17,18,20] All of these substitutions leave Cys 186 intact so that, despite the differences in carboxyl terminal sequences among membrane-bound proteins, there appears to be a common feature contained in the 180-189 sequence that allows membrane attachment, likely a structural feature common to these different sequences.

C. Monoclonal Antibodies Against P21 Protein and Domain Structure

A number of different monoclonal antibodies against P21 protein have been prepared.[21] These antibodies recognize different determinants. At present it appears that there are several different sets of determinants, each one corresponding to a different domain of the protein. One set of monoclonal antibodies against H-*ras* protein will not recognize other P21 proteins (such as N- and K-*ras*). Since these proteins all contain major differences only at their carboxyl terminal sequences, it is the carboxyl terminal end that is the determinant. Another set of monoclonal antibodies appears to be specific for position 12 while another set appears to be specific for a determinant around position 61.[21,22] Only the latter two sets inactivate the protein (GTP is not bound in the presence of the antibody).[21,22] Interestingly, it has been shown that microinjection of one of these monoclonal antibodies into transformed cells, temporarily causes them to revert back to phenotypically normal cells.[22] The results with each different set of monoclonal antibodies further support the idea of a domain structure for the P21 protein.

D. Implications for Structure-Function Relationships

From the above results, three major regions of the P21 proteins appear to be critical for function: the amino terminal segment containing the hydrophobic decapeptide Leu 6-Gly 15, the region containing Ala 59, Gln 61, and Glu 63, and the carboxyl terminal segment 180-189. Amino acid substitutions in each of these regions vitally affect protein function. The fact that the single amino acid substitutions of a wide variety of amino acids for Gly at position 12 and several different amino acids for Gln 61 cause malignant transformation of cells suggest that these substitutions cause changes in the structures of the peptide segments in which they occur. It is highly desirable to determine therefore what possible changes in conformation are induced by such substitutions and how they related to protein function. Determination of the manner in which amino acid substitutions alter protein structure is a special case of a general problem in protein chemistry, viz., the determination of the three-dimensional structure of a protein from its amino acid sequence. Theoretical and experimental methods exist and are being further developed to answer this question.[23-27]

II. METHODS AND APPLICATIONS

A. General Principle

The basic aim of the conformational energy calculations to be described below is to determine from amino acid sequence the most likely conformations of the transforming regions of the P21 proteins. It is known that for many, if not all, polypeptides and proteins, the linear sequence of amino acids that comprises the protein determines its unique three-dimensional (or tertiary) structure.[23-27] Considerable experimental evidence has accrued that strongly suggests that the observed structure(s) of a polypeptide or protein is (are) the one(s) of lowest free energy.[23-27] Computationally, it has been found that for macromolecules such as long polypeptide chains, proteins and protein-ligand complexes, the observed forms are those of lowest conformational energy. Determination of these lowest energy conformations or global minima requires solution to the multiple minimum problem. This latter term refers to the fact that a given polypeptide can be generated in numerous conformations that are of minimum energy locally. However, determination of the *lowest* energy form(s) of all of these local minima is the object of the calculations. Thus, in addition to computing the energy of any given conformation of a polypeptide it is necessary in principle to generate the possible conformations that the chain can adopt. The conformational energy of each such conformation is then computed and minimized to determine all of the local energy minima for the peptide. The energies of these minima are then compared to determine which structure(s) is (are) the one(s) of lowest energy, i.e., the global minimum or global minima. Each one of these two procedures used in the calculation of the structures of P21 peptide segments, computation of conformational energy and generation of the possible conformations of a given polypeptide, is now described in turn.

B. Computational Methods

1. Conformational Energy

The ground state or conformational energy of a given molecule, in particular of a polypeptide, can be computed according to the equation,[24-30]

$$E_{TOT} = \sum_{i\neq j} \frac{Q_iQ_j}{Dr_{ij}} + \sum_{i\neq j} \epsilon_{ij}\left[\left(\frac{R_{ij}}{r_{ij}}\right)^{12} - 2\left(\frac{R_{ij}}{r_{ij}}\right)^{M}\right] + \sum_{k} A_k(1 \pm \cos(n\Theta_k)) \qquad (1)$$

where E_{TOT} is the total conformational energy of the peptide; the first summation is the total coulombic electrostatic interaction energy between all interacting atoms i and j in the molecule; the second is the total Lennard-Jones interaction potential; and the third is the total torsional energy of the molecule. Q_i is the charge on the ith atom; D is the dielectric constant of the medium taken as 2; r_{ij} is the distance between the ith and the jth atoms; ϵ_{ij} is the potential well depth at the minimum in the Lennard-Jones potential function; R_{ij} is the distance between two atoms where the energy between them is a minimum; the exponent M is six for nonhydrogen-bonding atoms and ten for hydrogen-bonding atoms; A_k is the rotational barrier height for rotation around the kth bond; and n is two or three depending upon whether the bond is a single bond ($\cos 3\Theta_k$ or a double bond ($-\cos 2\Theta_k$); and Θ_k is the Kth dihedral angle.

A self-consistent and reliable set of coefficients for Equation 1 has been determined for all types of atoms that interact in a protein.[28-30] The first sum in Equation 1 is an electrostatic energy term summed over all varying pairwise interactions between charges on atoms. These charges have been computed by a reliable quantum mechanical method CNDO/2.[28-30] The second sum in Equation 1 is the nonbonded interaction energy summed over all varying pairwise interatomic interactions. This energy is due to attractive induced dipole-induced dipole interactions that depend on the atomic polarizabilities of the interacting atoms and vary as the inverse sixth power (tenth power for hydrogen bonding atoms) of the distance between the pair of interacting atoms and on the repulsion between two interacting atoms, should the distance between them become less than the sum of their van der Waal's radii.[28-30] The constants ϵ and R in Equation 1 have been parameterized from a large body of experimental cyrstal and gas phase data[28] so that they reproduce the cell lattice constants for crystals of small, uncharged molecules. These constants are fully transferable from one crystal structure to another, and the sum of the first two terms in Equation 1 can reliably compute the sublimation energies of crystals where they have been determined experimentally.[28]

The last term in Equation 1 is the torsional energy of the molecule[29,30] that depends on the values of the dihedral angles and hence on the particular bonds about which rotation takes place, such as carbon-carbon single bonds, C–N peptide bonds, etc., the constants for which have been determined experimentally.[28-30] The potential functions in Equation 1 have been used for a variety of structural problems to be described and used to develop long-range attractive potentials for amino acid residues that save much computational time.[31]

It should be noted that in the calculations performed using this approach, only the dihedral angles vary; the bond lengths and bond angles remain fixed. This constraint is based upon the observations[24,25,28-30] that the bond lengths and bond angles for different amino acid residues remain remarkably constant regardless of their neighbors or the crystal environment. This constraint simplifies energy computations considerably.

Generation of the cartesian coordinates of the atoms of a polypeptide from specified dihedral angles and computation of the energy of this conformation according to Equation 1 are accomplished using the program ECEPP (Empirical Conformation Energies for Peptides and Proteins).[29,30]

a. Solvation

Computation of the low energy conformations for some polypeptide chains includes the effect of aqueous solvent on the conformational preferences of these peptides. Segments such as Ile 55-Met 67 in the P21 protein (whose conformational preferences are discussed below) may be exposed to the solvent, water, thereby necessitating its inclusion in the calculations. A recent treatment of the effects of water on the conformations of amino acid and simple peptides[32,33] considers each amino acid residue as containing two types of water: nonspecific and specific water of hydration.

1. Nonspecific Hydration

The nonspecific hydration refers to the hydration volume available to each atom in the amino acid (or end group) residue. This maximal volume of hydration has been computed by packing the maximal number of water molecules around a central atom that will result in low energy configurations.[32,33] If a given atom, the central atom, is approached by other atoms (that are not covalently linked to the central atom), the volumes of hydration may interpenetrate. There is thus a *decrease* in the hydration volume of the central atom. The energy for this decrease has been expressed as,[32]

$$E_{hy,i} = \frac{V_{H,i} - V_{ex,i}}{V_{H,i}} \cdot E_{h,i} \qquad (2)$$

where $E_{hy,i}$ is the nonspecific hydration energy for central atom i, $E_{H,i}$ is the energy of hydration for a specific group such as -CH_2, amide nitrogen, etc. The values of $E_{H,i}$ are taken from experimentally determined values for small molecules.[32] Positive values for $E_{H,i}$ (as for aliphatic groups) indicate *unfavorable* solvation energy while negative values indicate favorable solvation energies. $V_{H,i}$ is the unoccupied hydration volume of the central atom, and $V_{ex,i}$ is the hydration volume of the central atom occupied by another noncovalently bound atom.

2. Specific Hydration

The specific hydration refers to specific water molecules that are hydrogen-bonded to polar atoms (N, O, and the hydrogens attached to them) for both backbone and side chain. It is assumed that there is one specifically bound water molecule per polar atom. The interaction energy between the specifically bound water molecule and the central atom to which it is bound is given by a hydrogen-bonding energy reliably estimated at −1.0 kcal/mol.[32] The energy of specific hydration for a polar central atom i is given by the expression,[32]

$$\text{Esp,i} = \text{Espec,i}(1 - b) \qquad (3)$$

where Esp, i is the conformation-dependent specific hydration energy, Espec, i is the specific hydration energy of central atom i with no displacement of the specifically bound water by other atoms, and b is a number varying from 0 to 1 and depends directly on the degree of overlap between the van der Waal's radius of other atoms and the specific hydration radius of the central atom.[32] As overlaps increase, b approaches 1, and the specific hydration energy approaches 0. With no overlaps, the value of b is 0 and Esp,i = Espec,i (−1.0 kcal/mol for specific hydration). The total hydration energy for a given peptide is then given by

$$E_{HYD} = \sum_{N} E_{hy,i} + \sum_{M} E_{sp,j} \qquad (4)$$

where the index limit N is the total number of atoms in the molecule, M is the total number of polar atoms in the molecule, and the second sum in Equation 4 is taken over only the polar atoms.

The hydration energy computed here is that of a first solvent shell around the molecule in which water molecules pack around individual atoms and hydrogen bond to polar atoms. It does not take into account the effects of "outer shell" water molecules or entropic effects on water structure, although entropic effects are included implicitly in the values of $E_{h,i}$.[32] It does, however, include the specific conformational effects of water on polypeptide structure. This first shell effect appears to be the major effect of solvent on polypeptide structure.[32] In addition, because of the unfavorable $E_{H,i}$ values for the aliphatic side chain atoms of nonpolar residues, there is a tendency for nonpolar atoms to pack and exclude bound water, a "hydrophobic" effect.

This solvation treatment has been incorporated into the basic approach as in Equation 1[32,33] and has been employed to recompute the single residue minima of the amino acids[32] and dipeptides.[32] The single residue minima determined with solvation are quite similar to those from the vacuum calculations[32] although higher energy forms computed by the latter method are lower in energy with solvation. The solvation treatment of dipeptides has enabled calculations of bend probabilities for different known bend-forming dipeptides exposed to solvent. With solvation, predicted bend probabilites for specific dipeptides agree closely with actually observed frequencies of these dipeptides in bends in proteins,[32] an agreement not obtained with *in vacuo* calculations.[32] Evidence therefore exists that these solvation parameters are reliable in structure calculations.

2. Generation of Possible Structures

a. Single Amino Acid Residue Minima

The low energy conformations of all of the 20 naturally occurring blocked amino acid residues,[34,35] substituted amino acids,[34,35] and saccharide residues[36] have been determined. In the case of the naturally occurring amino acid residues, all of the allowed conformations for the simplest amino acids such as Gly and Ala have been determined by generating all combinations of backbone dihedral angles PHI (rotation around N–C^{α} bond) and PSI (rotation around C^{α}–C′ bond) for small (5°) increments of each. Computation of the energies of each conformation allows identification of values of these dihedral angles where local minima occur. These conformations are subjected further to energy minimization to determine the exact values of the backbone dihedral angles at these minima,[34,35] referred to as single residue minima.[34,35] The single residue minima for all other amino acids have been determined by building up the positions of side chain conformations on the basis of the results for L-Ala as described in detail elsewhere.[34,35]

b. Use of Single Residue Minima to Build Oligopeptide Chains

Generation of all possible conformations of even an oligopeptide chain would result in an astronomically large number of structures for each of which the energy must be computed and then minimized. Instead of this feasible approach to solving the multiple minimum problem alluded to above, use is made of a systematic chain build-up procedure in which amino acid residues are added to one another sequentially in all of their allowed conformations.[24-27]

In this method, the single residue for the first two amino acid residues in a given sequence is combined, and all of the resulting conformations subjected to energy minimization. The energy minimization routine employs an efficient line-search algorithm.[37] In all minimizations, all of the variables, i.e., the dihedral angles of the polypeptide, are varied. The algorithm has been modified so that the dihedral angles that will cause the greatest decrease in energy are changed first followed by the remaining variables.[38] For each starting conformation generated by the combination procedure described above, one minimum energy conformation results. All of the resulting low energy conformations whose conformational energies lie within a cutoff value of that of the global minimum are retained.

The minimum cutoff value used is 3 kcal/mol which has been shown to include all possible contributing structures at the dipeptide level.[24-27] For longer sequences, up to five residues, the cutoff value depends upon the particular amino acids in the sequence. For polar-residues, an additional 3 kcal/mol is added to the basic 3 kcal/mol to include possible additional hydrogen bonding effects that may occur at some further stage of the chain buildup procedure.[38]

All possible combinations of these resulting low energy dipeptide conformations with the single residue minima for the next amino acid residue in the sequence are generated, and the resulting tripeptide conformations subjected to energy minimization, and the procedures described above are then repeated. This process is repeated until the low energy conformations of the desired peptide are generated. Thus, the lowest energy conformations for a given peptide are generated using a "directed search" method, i.e., one in which the search for preferd conformations is "directed" by the use of low energy conformers of amino acid residues and short peptide fragments.

C. Applications of Methods to Oligopeptides and Protein-Ligand Complexes

This procedure has been successful in computing the conformations of oligopeptides (containing up to six or seven residues)[24,25] and oligosaccharides[36] with good agreement between the computed structures and those obtained experimentally, e.g., from X-ray crystallography and/or from solution NMR studies. In several cases, this chain buildup procedure has been employed in calculation of the structures of polypeptides that are constrained, in particular gramicidin S, a cyclic decapeptide[39] and collagen models such as triple-stranded (Gly-Pro-Pro)$_{10}$.[40] These cases are now discussed.

1. Gramicidin S

This cyclic decapeptide antibiotic has the sequence:

L-Val-L-Orn-LLeu-D-Phe-l-Pro
L-Pro-D-Phe-L-Leu-L-Orn-L-Val

The number of allowed conformations of this polypeptide is limited because the sequence is repetitive and the peptide is cyclic. The cyclic property imposes the stringent constraint that all structures generated for the molecule must satisfy a ring-closure condition.[39] Application of the chain buildup procedures described above used in conjunction with the ring-closure condition nontheless required the minimization of the energies of over 10,000 starting conformations.[39] Of the resulting energy-minimized conformations, only one lowest-energy or global-minimum-energy structure was obtained; it was approximately 2 kcal/mol lower in energy than the next lowest energy structure.[39] Calculations of the expected nuclear magnetic resonance (NMR) ^1H coupling constants for side chains and backbone vicinal proton pairs were in excellent agreement with available NMR data for this molecule.[39] Subsequently, the structure of gramidicin S was determined by X-ray crystallography[41] and found to be quite similar to the calculated one.[39] Raman and infrared (IR) spectral studies, interpreted with the aid of normal mode calculations also support this structure.[42] In both structures (computed and X-ray), there is an extensive network of intramolecular NH- - -O=C hydrogen bonds across the chain between the two Val-Orn-Leu sequences which exist in an antiparallel β-pleated sheet conformation with respect to one another. The two Phe-Pro sequences form β-bends. In addition, in each structure, an important hydrogen bond forms between the δ-NH_2 group of Orn (not one of the naturally occurring amino acids but one for which the single-residue minima has been determined[34]) and the C=O of a preceding D-Phe residue. Thus, specific hydrogen bonds may well stabilize the structure of this molecule. Possible substitution of L-amino acids for D-Phe and amino acids other than Lys for Orn in this peptide might be expected to destabilize this structure and possibly alter its function.

2. Collagen

Collagen is a fibrous protein found in connective tissue. It has been shown to be composed of triple helixes formed by three strands of single polypeptide chains whose sequences are of the form poly(Gly-X-Y).[43] The chain buildup procedure described above was thus applied to computation of the preferred conformations for poly(Gly-Pro-Pro).[40] The calculations in this case were aided by the constraint of the repetitive sequence and the limited number of conformations (single residue minima) that can be adopted by Pro.[34,35] Initially, all of the low-energy minima for Gly-Pro-Pro were determined (the conformational space of this unit being constrained by the few preferred conformations for Pro). The allowed conformations for this basic unit could then be combined to generate the preferred conformations for any desired chain length. For a single chain of $(Gly\text{-}Pro\text{-}Pro)_4$, there were numerous low-energy conformations. However, when each of these was packed into triple-stranded helixes satisfying symmetry constraints (assuming that the three chains are equivalent), remarkably only one lowest-energy structure (helix) was found, which was 19 kcal/mol lower in energy than the structure of next-lowest energy.[40] After these calculations were completed, the X-ray crystal structure of $(Gly\text{-}Pro\text{-}Pro)_{10}$ was determined.[44] Comparison of the atomic coordinates of the computed global minimum-energy structure with those of the X-ray structure revealed a root-mean-square deviation of only 0.3 between them, i.e., the structures were superimpossible.[40,44] The three chains pack in a way that they coil around one another thereby forming ''coiled coils''. Hydrogen bonds form between the NH of the Gly residue of one chain and the C=O of the Pro residue of a neighboring chain, greatly stabilizing both the intrachain conformation and the packing arrangement.[40] Similar calculations have been performed for Poly(Gly-Pro-Ala)[45] which has been computed to adopt the collagen-like ''coiled coil'' conformation found for Poly(Gly-Pro-Pro).[40,44] However, poly(Gly-Ala-Pro)[46] does not adopt this ''coiled coil'' structure as its lowest energy conformation, a demonstration of the sensitivity of structure to specific sequence.[46]

The structure-stabilizing long-range interactions that exist between chains in collagen result from the tight packing of chains (involving many low energy weak nonbonded interactions described by the second sum in Equation 1, but ones which add up to produce significant stabilization energy) and from specific hydrogen bonds. These interactions result in selection of only one conformation from many available ones for the single chain. Such a process may well be involved in the folding of globular proteins where two or more sequences of the same chain interact specifically so that only one of many possible conformations for each sequence is selected.

Thus, application of the buildup procedure for constrained polypeptides, as described above, results in predictions of structures that are in agreement with experiment. The method can further be extended to incorporate multiple interchain interactions as in the case of collagen. In fact, a modification of this procedure has been applied to determination of the three-dimensional structures of enzyme-substrate complexes[47] in the cases of chymotrypsin and lysozyme.[47,48,50-53] The results obtained in the case of lysozyme are briefly reviewed here to illustrate that the method can be applied to computation of the three-dimensional structures of bimolecular complexes (such as are involved in GTP-P21 protein complexes), and ones that involve interactions between proteins and nonprotein polymers, in this case polysaccharides.

3. *Enzyme-Substrate Complexes*

Lysozyme (muramidase), in this case hen egg white or chicken lysozyme is an enzyme with 129 amino acid residues that cleaves bacterial cell walls composed of alternating units of *N*-acetylglucosamine (GlcNAc) and *N*-acetylmuramic acid (MurNAc) in β-1, 4-linkage.[54] It also cleaves chitin (poly-β-GlcNAc) with high efficiency.[54] It is known that at most six saccharide units can bind to the enzyme at once. These sites are labeled as subsites A-F (from nonreducing to reducing end). Cleavage occurs between sites D and E.[54] In the case of hexa-GlcNAc or $(GlcNAc)_6$, cleavage results in predominant formation of $(GlcNAc)_4$ and $(GlcNAc)_2$ although some other oligosaccharides also form in smaller amounts. Catalysis is performed by two residues between sites D and E, Asp 52, and Glu 35.[54] The X-ray crystal structure of $(GlcNAc)_3$ bound to lysozyme in sites A-C was determined.[54] Three more residues were added onto this structure by model-building to explore how substrates bind in the active site.[54] Because of bad contacts between the fourth residue from the nonreducing end of $(GlcNAc)_6$ and some of the enzyme amino acid residues in site D, it was necessary to distort the sugar ring from a chair form to a half-chair form.[54] In this form, the D-sugar residue binds deeply in the active site cleft between Asp 52 and Glu 35. While of significantly higher energy than a chair form, the half-chair form resembles the structure that a carbonium ion would adopt in the catalytic mechanism, and strong evidence exists that this mechanism occurs in the lysozyme-catalytic reaction.[54] Thus, ground-state "strain" would contribute to lowering the activation energy.[54] Further model-building studies suggested that two more saccharide units could be accommodated in sites E and F. In the latter site, Phe 34 and Arg 114 make favorable contacts with the reducing saccharide residue on the "right side" of the active site.[47,48,50-53]

Conformational energy calculations were applied to the problem of determining the structures of enzyme-substrate complexes of lysozyme in a systematic five-stage procedure.[50] First, using techniques completely analogous to those described above for polypeptides, the allowed conformations of poly-GlcNAc and poly-(GlcNAc-MurNAc) were determined.[36] The results compared quite favorably with available experimental data on such molecules as cellulose and chitin (fiber diffraction patterns).[36] Second, the X-ray crystal coordinates for lysozyme were subjected to energy refinement so as to remove a number of bad contacts and to optimize a number of favorable interactions (such as hydogen bonds). Third, a molecular fragment of $(GlcNAc)_2$ with no side chains was employed in an eight-dimensional (six rigid body and two inter-ring dihedral angle variables) nonbonded energy map of the entire active site and surrounding regions to determine the allowed binding regions in and around the active site.[52] Fourth, the full molecule $(GlcNAc)_2$ was placed in all computed allowed conformations found in Step 1 in the regions found in Step 2. These structures were then subjected to energy minimization.[52,53] Fifth, the lowest energy structures from Step 4 were then used as starting conformations to which sugar residues were added to either the reducing or nonreducing end in all of their allowed conformations as determined in Step 1.[51-53]

It was found that a maximum of six GlcNAc saccharide residues bound to the active site in agreement with experiment.[50,53,54] The most stable (lowest energy) conformation for $(GlcNAc)_6$ bound to lysozyme was found to bind in such a way that the sugar residues in sites A-C adopted a conformation identical to that in the X-ray crystallographic structure of $(GlcNAc)_3$ bound to lysozyme.[47,48,50-53] The fourth residue in site D, however, was found to bind near the surface of the active site cleft in the *full chair* form and *not* in the distorted half chair form.[47,48,50-53] In addition, the residue in the F site was found to bind to the "left side" of the active cleft involving such residues as Arg 45, Asn 46, and Thr 47 as opposed to the "right side" of the cleft involving such residues as Phe 34, Asn 113, and Arg 114,[47,48,50-53] as found in the model-built structure (see Figure 1). Thus, the calculation resulted in predicted structures for *inhibitors* of lysozyme such as $(GlcNAc)_3$ in agreement with experiment. However, the calculations resulted in predicted structures for *substrates* of lysozyme such as $(GlcNAc)_6$ that differed from the results of model-building studies.

At the time that the above calculations were being completed, it was found independently that Gd^{2+} ion that binds deeply in the active site cleft between the two catalytic residues Asp 52 and Glu 35 does not alter the affinity of $(GlcNAc)_4$ for lysozyme as it should if $(GlcNAc)_4$ binds deeply in the cleft.[55] Furthermore, an X-ray crystallographic experiment was performed subsequently in which the inhibitor MurNAc-GlcNAc-MurNAc was bound to sites B-C-D of lysozyme, the first crystallographic structure in which there was 100% occupancy of the D site by a saccharide residue.[56] The conformation of the MurNAc sugar residue in site D was virtually identical to the predicted structure of the GlcNAc sugar residue in site D in the computed structures for $(GlcNAc)_n$, n = 4, 5, 6 and for copolymers of GlcNAc and MurNAc.[47,48,51] In the X-ray structure, the D sugar residue was bound near the surface of the active site cleft in the *full chair* form as predicted.[56]

Further experiments have recently been performed to determine the position of binding of the last two residues of $(GlcNAc)_6$ in sites E and F.[57] In one experiment, the binding affinities of $(GlcNAc)_6$ to hen egg white lysozyme and pheasant lysozyme were measured.[57]

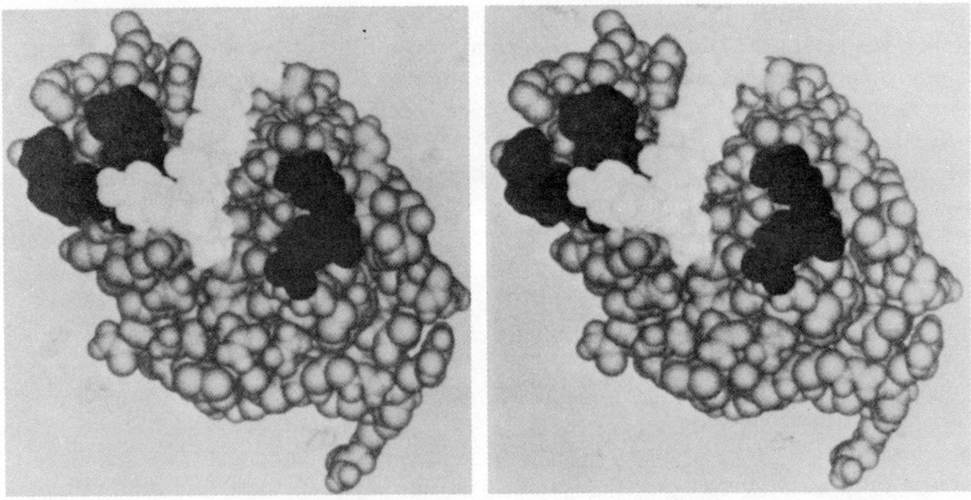

FIGURE 1. Space-filling stereo view of the "left-sided" binding conformation of $(GlcNAc)_6$ to lysozyme. The residues of the enzyme involved in both left- and right-sided binding are shaded dark gray.

These two enzymes have identical active sites except that Lys and His in pheasant lysozyme replace Asn and Arg at positions 113 and 114, i.e., the critically important residues on the "right side" of the active site of hen lysozyme are replaced nonconservatively. The binding affinity of the hexasaccharide to each enzymes was the same, casting doubt on the role of the "right-sided" residues in ground-state binding.[57] In a second experiment, the ability of different oligomers of GlcNAc, to displace a monoclonal antibody to the "left-sided" binding site of lysozyme was measured. This monclonal antibody has been shown to bind to the "left side" residues in such a way that it does not extend into the active site cleft.[57,58] In cryoscopic experiments,[57] it was shown that short saccharides that bind only in sites A-D such as GlcNAc, $(GlcNAc)_2$, and $(GlcNAc)_4$ do not displace the antibody while substrates, especially $(GlcNAc)_6$, known to bind to site F, significantly displace the monoclonal antibody (approximately 50% at concentrations around the k_m). Thus, the existence of a "left-sided" binding site was directly confirmed in solution studies.

A stereo space-filling model of the computed "left-sided" binding mode of $(GlcNAc)_6$ to lysozyme is shown in Figure 1. In this figure the "left-sided" binding residues and proposed "right-sided" binding residues are shaded in black. The chain buildup methods have thus been successful in predicting the structures not only of constrained polypeptides but also of macromolecular enzyme-substrate complexes where the structure of the active site is known. It was therefore desirable to extend the above-described methods to long unconstrained polypeptides and proteins. The buildup procedures, however, when extended to nonregular sequences that have no constraints, generate very large numbers of low energy conformations for short peptide segments making repeated application of the method infeasible.

D. Extension of the Buildup Methods to Long, Unconstrained Polypeptide Chains

Recently, a method has been devised that circumvents this problem.[38,59] When the low energy structures for oligopeptides such as tri- and tetrapeptides are examined, it is found that many of the conformations are the same, i.e., the backbone dihedral angles are close in value while the side chain dihedral angles may adopt different rotational isomeric states. If only minima with distinctly different backbone conformations are considered, the number of conformations for a given oligopeptide is reduced greatly, thereby making feasible the buildup of long polypeptide chains. This method, called the method of combination of nondegenerate minima,[38,59] retains all possible representative low energy conformations of peptide chains while it eliminates excessive numbers of degenerate backbone states. This method has been tested and recently applied to the problem of computing the structures of membrane polypeptides and membrane-active proteins.[38,59] These polypeptides are selected because they are known to require nonaqueous environments to fold properly. Thus, intramolecular interactions would be expected to determine the three-dimensional structure, and solvent effects would not be expected to contribute significantly to determination of this structure. In cases where polar residues are involved and solvation effects may become important, adaptations of the above method are employed as described in the solvation section.

1. Applications of Methods to Long Polypeptides

a. Leader Sequences

The above-described methods have been employed in computing the preferred conformations for membrane-bound leader sequences[59] and for the membrane-active protein melittin.[38] In the case of the leader sequence for murine pre-κ light chain, whose sequence is Met-Glu-Thr-Asp-Thr-[(Leu)$_3$-Trp-Val]$_2$-Pro-Gly, the preferred conformations for the hydrophobic decapeptide [(Leu)$_3$-Trp-Val]$_2$ were determined, and it was found that of numerous possible structures there is one unique low energy structure for this sequence, an α-helix, although the terminal Val residue can adopt a number of non-α-helical conformations at the end of this helix.[59] When the remaining residues were added in all of their single residue minima to the growing polypeptide chain, a number of low energy structures resulted in all of which the residues from Leu 6-Trp 15 were in the α-helical conformation. This α-helix is terminated by a kink in the chain at the terminal Val-Pro-Gly that induces a reversal in the chain-direction.[59] The initial polar sequence Glu 2-Thr 5 exists in a number of β-like conformations, the Asp-Thr dipeptide adopting C-7 conformations[59] (see Reference 34 for definitions of conformational states). The Pro residue causes helix termination because amino acid residues preceding Pro cannot adopt the A (α-helical) state without making bad contacts with some of the Pro ring atoms.[60] This structure has been compared with available physical, chemical, and genetic experimental data on leader sequences.[61-65] A number of CD studies[61] on the conformations of leader sequences show high (approximately 50%) α-helix contents in nonaqueous solvents. (The computed structure is 50% α-helical).[59] In addition, genetic experiments in which β-hydroxyleucine, a known α-helix breaker incorporated in place of leucine into a growing leader sequence in a cell-free synthetic system causes greatly diminished protein translocation.[62] Leader sequences with deletion mutations of hydrophobic residues that result in the separating of two Pro residues by only several amino acids likewise cause highly impaired protein translocation.[63] The approximation of two Pro residues in a sequence is expected to cause termination of an α-helix.[53] Other data suggest that the long stretches on nonpolar residues that occur in a wide variety of leader sequences are α-helixes that end in hairpin turns that may be "signal" conformations recognized by membrane-bound peptidases.[64] Amino acid sequencing studies of leader peptides indicate that the initial (computed) less structured polar segments found in many sequences may not be necessary for proper functioning.[65]

b. Melittin

This membrane-active protein which is present in bee venom and causes rupture of cell membranes is composed of 26 amino acid residues and requires micelles or a nonpolar solvent to fold.[66] The sequence is

Gly-Ile-Gly-Ala-Val-Leu-Lys-Val-Leu-Thr- 10
Thr-Gly-Leu-Pro-Ala-Leu-Ile-Ser-Trp-Ile- 20
Lys-Arg-Lys-Arg-Gln-Gln-NH_2

Table 1
LOW ENERGY CONFORMATIONS OF *N*-ACETYL-VAL-LEU-LYS-VAL-LEU-NHCH$_3$

	Conformational state					
Conformer	**Val**	**Leu**	**Lys**	**Val**	**Leu**	**Relative energy (kcal/mol)**
1	A	A	A	A	A	0.0
2	A	A	A	A	C	0.3
3	C	A*	G	A	C	1.4
4	A	C	G	A	A	1.4
5	C	A*	G	A	C	1.9
6	A	C	G	A	C	2.1
7	A	A	A	C	C	2.5
8	C	A	C	A	D	2.6
9	C	D	D	C	A	2.7
10	C	D	D	C	D	2.8

Note: Conformational states are defined as follows: states in which dihedral angles are represented by a single-letter code are defined in Zimmerman et al., *Macromolecules,* 10, 1, 1977. The familiar states are A(α-helix) and E(extended). States, E, E*, F, F*, D, and D* are collectively called "beta" conformations. The actual dihedral angle ranges for all single-letter states are A, $-110° \leq \phi < -40°$, $-90 \leq \psi < -10°$; C, $-110° \leq \phi < -40°$, $50° \leq \psi < 130°$; D, $-180° \leq \phi < -110°$, $20° \leq \psi < 110°$; E, $-110° \leq \phi < -40°$, $-180° \leq \psi < -140°$ or $130° \leq \psi < 180°$; F, $-110° \leq \phi < -40°$, $-180° \leq \psi < -140°$ or $130° \leq \psi < 180°$; G, $-180° \leq \phi < -110°$, $-90° \leq \psi < -40°$. States indicated by asterisks are obtained by multiplying the corresponding single-letter state by -1 reversing the inequalities. All energies are relative to the energy of Conformer 1. In this table all conformers whose energies lie within 3 kcal/mol of the energy of Conformer 1 are included. In all subsequent tables, only those conformers whose energies lie within 2 kcal/mol of the energy of the global minimum are included.

The allowed structures for the membrane-bound portion of this molecule, Gly 1-Ile 20 have been computed.[38] First, the preferred conformations were explored for the two long hydrophobic sequences 1-9 and 13-20, both of which show distinct preferences to adopt the α-helical conformation.

For example, in the case of the pentapeptide Val-Leu-Lys-Val-Leu (residues 5-9 in the sequence) from approximately 2000 starting conformations, only ten structures were found that lay within the cutoff of 3 kcal/mol, the lowest in energy being α-helical, as shown in Table 1.[38] It may be observed in this table that several of the low energy structures (such as conformer 10) are not helical. Thus, while this pentapeptide is conformationally restricted, experimental methods such as circular dichroism designed to determine secondary structure would detect a "random coil" for this peptide when in fact the number of actual structures adopted by the peptide is quite limited. In contrast, energy minimization of about 1000 starting conformations of Thy-Thr-Gly-Leu-Pro (residues 10-14 in the sequence) resulted in approximately 50 different low energy structures. This result indicates that residues 5-9 tend to form a nucleation site in the chain while the segment containing residues 10-14 is much less structured. When the allowed structures for residues 5-9 were combined with those for Thr-Thr-Gly (residues 10-12), only three structures were obtained whose energies lay within a cutoff of 10 kcal/mol of the energy of the global minimum.[38] Further combination of these structures with those found for the segment 13-20 resulted in only two structures

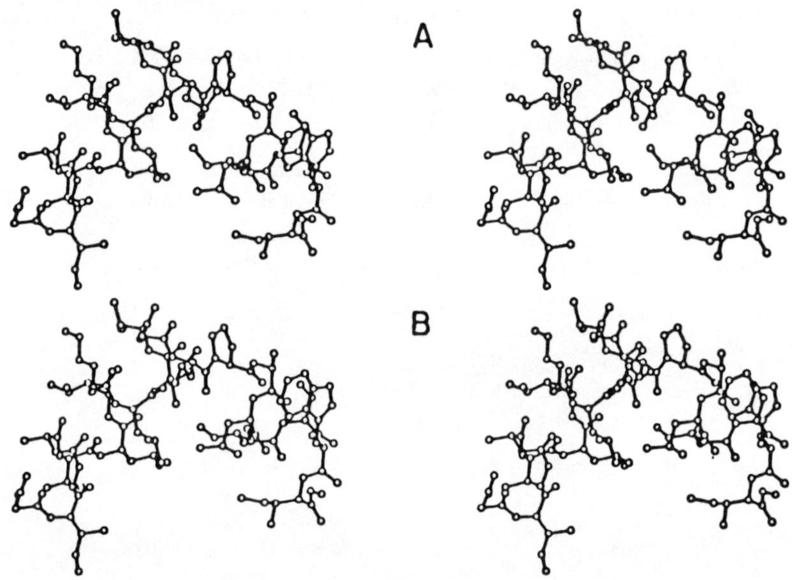

FIGURE 2. Stereo views of the two computed lowest energy structures for melittin (A and B) which are seen to be quite similar to one another. The amino terminus for both structures is on the left of the figure.

for the whole sequence. Both of these structures were quite similar to one another. Both were compact structures in which segments 1-10 and 14-10 formed α-helixes separated by a bend at residues 11-13. The axes of each helical segment made an angle of approximately 100° with one another. Both structures are shown in stereo view in Figure 2.

The overall structure of the molecule is such that the side chains of nonpolar residues pack in the inner surface of the molecule while the polar side chains point toward the outer surface of the molecule, as shown in Figure 2. These features computed for the molecule are in good agreement with experimental structural studies on the conformation of melittin[66,67] in which the molecule was found to be a bent α-helical rod in which the axes of the two helical segments made an angle of about 100° and in which a turn occurred at Thr 11-Gly 12. The nonpolar residues packed in the interior of the molecule while the side chains of the polar ones pointed toward the external surface of the molecule as in Figure 2.

The agreement between computed structures and experimental results for a wide variety of cases, i.e., oligopeptides, protein-ligand complexes and polypeptides, and membrane proteins, indicates that the methods are reliable and can be used to explore the preferred conformations for the polypeptide sequences whose structures are unknown, in particular those that occur in the P21 proteins, whose structures are now discussed.

III. STRUCTURES OF THE TRANSFORMING REGIONS OF THE P21 PROTEINS: RESULTS AND DISCUSSION

In the case of the P21 proteins, the results of computations on the structures of different segments of the protein known to be involved in cell transformation are compared with the results of genetic experiments. Thus, in addition to structure analysis, the structure of these peptides is correlated closely with function, i.e., ability of the protein to cause cell transformation.

As noted in the Introduction, transforming activity of the P21 protein is clustered in three discrete segments: 4-21, the segment containing residues 59, 61, and 63 and the carboxyl terminal decapeptide including residues 181-189. The results obtained on each of these segments and the implications for cell transformation are now discussed.

Table 2
LOW ENERGY CONFORMATIONS FOR THE HYDROPHOBIC DECAPEPTIDE IN THE "NORMAL" PROTEIN *N*-ACETYL-LEU-(VAL)$_3$-GLY-ALA-GLY-GLY-VAL-GLY-NHCH$_3$

	Conformational state										
Conformer	Leu	Val	Val	Val	Gly	Ala	Gly-12	Gly	Val	Gly	Energy (kcal/mol)
1	A	A	A	A	A	C	D*	A	A	C	0.0
2	A	A	A	A	A	A	C	D*	A	A	0.8
3	A	A	A	E	D*	D	D*	A	C	F*	0.9
4	A	A	A	A	A	D	D	D*	A	A*	1.2
5	A	A	A	A	A	A	C	D*	A	A*	1.2
6	A	A	A	A	A	C	D*	A	A	A*	1.6
7	A	A	A	E	D*	D	D*	A	C	D*	1.7
8	A	A	A	A	A	C	D*	A	A	D*	1.7
9	A	A	A	A	A	C	D*	A	A	A	1.9
10	A	A	A	A	A	A	C	D*	A	D	1.9
11	A	A	A	E	D*	D	D*	A	C	E*	2.0
12	A	A	A	A	A	A	A	A	A	A	2.0

Note: See legend to Table 1 for explanation.

A. Structures of the Hydrophobic Decapeptide Leu 6-Gly 15

The first known critical single substitution that alters the function of the P21 protein is Val for Gly at position 12, a substitution which occurs in the middle of the hydrophobic decapeptide Leu 6-Gly 15. The allowed conformations for this peptide have been computed with both Gly and Val at position 12. The allowed conformations for the hexapeptide, Leu-Val-Val-Val-Gly-Ala (residues 6-11) were first determined. The lowest energy structure for this segment was an α-helix. However, a number of other low energy structures were obtained that were distinctly nonhelical.[68] The single residue minima for either Gly or Val were then combined with the low energy hexapeptide conformations and the resulting structures subjected to energy minimization. Each remaining amino acid was added to the "growing-chain" in an identical manner until the low energy structures for the entire decapeptide were generated,[68,69] with either Gly or Val at position 12.

The results are shown in Tables 2 (peptide with Gly at position 2) and 3 (peptide with Val at position 12).[68] It may be noted from comparison of Tables 2 and 3 that the peptide with Gly 12 is flexible, i.e., there are a number of different low energy conformations allowed for it, while the peptide with Val 12 is more constrained. Furthermore, the Gly 12-peptide contains significant nonhelical structure, while the Val 12-peptide is more α-helical.

Two prominent structural features of the Gly 12-peptide (Table 2) are that the first three-residues (Leu 6-Val 8) are α-helical and the residues 11-12 tend to adopt the CD* conformation (see legend to Table 1 and Reference 34 for definitions). The latter structural feature is discussed below.

Comparison of the global minima of both peptides reveals that they contain an α-helix that terminates in a chain reversal (CD* conformation) (see footnote to Table 1 for definition of conformational states). This chain reversal occurs at Ala 11-Gly 12 in the Gly 12-peptide and at Val 12-Gly 13 in the Val 12-peptide. From Tables 2 and 3, it may be seen that Conformer 2 for the Gly 12-peptide (Table 2) is the same as Conformer 1 for the Val 12-peptide (Table 3). No conformation exists for the Val 12-peptide, however, that is the same as Conformer 1 of the Gly 12-peptide. The reason for this latter result is that in the global minimum structure for the Gly 12-peptide (Conformer 1, Table 2) and in many of the higher

Table 3
LOW ENERGY CONFORMATIONS FOR THE HYDROPHOBIC DECAPEPTIDE IN THE TRANSFORMING PROTEIN *N*-ACETYL-LEU-(VAL)$_3$-GLY-ALA-VAL-GLY-VAL-GLY-NHCH$_3$

	Conformational state										
Conformer	Leu	Val	Val	Val	Gly	Ala	Val-12	Gly	Val	Gly	Energy (kcal/mol)
1	A	A	A	A	A	A	C	D*	A	A	0.0
2	A	A	A	A	A	A	C	D*	A	A*	1.0
3	A	A	A	A	A	A	C	D*	A	D*	1.9
4	A	A	A	C	D*	A	A	A	A	A	2.0

Note: See legend to Table 1.

energy structures, Gly 12 exists in the D* state ($\phi = 170°$, $\psi = -60°$), a conformation that is energetically forbidden for the Val 12-peptide. Conformer 1 in Tables 1 and 2 are compared in stereo view in Figure 3. The Gly 12-peptide (speckled circles) and Val 12-peptide (open circles) are different mainly in the area of the arrows the Cα'S of residue 12 of the two peptides. Thus the substitution of Val for Gly causes a significant difference in structure in the P21 hydrophobic decapeptides.

These results suggest that no L-amino acid can substitute for Gly at position 12 without major structural alteration and change in function. This conclusion has firm experimental support. It is known that there are P21 proteins with substitutions of amino acid residues such as Lys, Ser, Arg,[1,2] and Asp[5] for Gly at position 12. All of these proteins cause malignant transformation of NIH 3T3 cells. Recently, genetic experiments have been performed in which site-specific mutations in the *ras*-gene have been made at only the 12th coding oncogene triplet such that all of the naturally occurring L-amino acids have been substituted for Gly at position 12.[14] All of these proteins except the one containing Pro at position 12 caused malignant transformation of cells.[14]

The fact that the Gly 12-peptide can adopt a structure identical to the global minimum energy structure of the Val 12-peptide further suggests that elevated intracellular concentrations of the normal protein (containing Gly 12) can also produce malignant transformation of cells. At sufficiently elevated concentrations of the normal protein, the higher energy form of the Gly 12-peptide (Conformer 2 of Table 2) may be sufficiently elevated to cause the expression of a malignant phenotype. This conclusion is corroborated by the recent discovery that the proto-oncogene, when spliced onto a viral LTR (long terminal repeat sequence that causes unrestricted expression of a neighboring gene) causes malignant transformation of NIH 3T3 cells.[11] The proto-oncogene itself does not cause malignant transformation of cells into which it has been transfected[1,2] as noted in the Introduction. The elevated intracellular protein concentration is therefore responsible for cell transformation as inferred from the calculations, a result also corroborated in microinjection experiments.[4]

As mentioned in the Introduction, it has been found that the synthetic P21 peptide Try 4-Ile 21 has been synthesized and found to bind with high affinity to GTP. We have recently obtained the one- and two-dimensional NMR spectra of this peptide with either Gly or Val at position 12.[77] All residue assignments have been made. Comparison of the amide proton spectra of both peptides reveals that each peptide adopts a significantly different conformation. The chemical shifts of the aliphatic protons appear to occur at very similar field strengths, with the exception, of course, of the Gly 12 and Val 12 protons. Of considerable interest is the fact that the C$^{\beta}$H of one of two of the Ala residues (11 and 18) of the Val

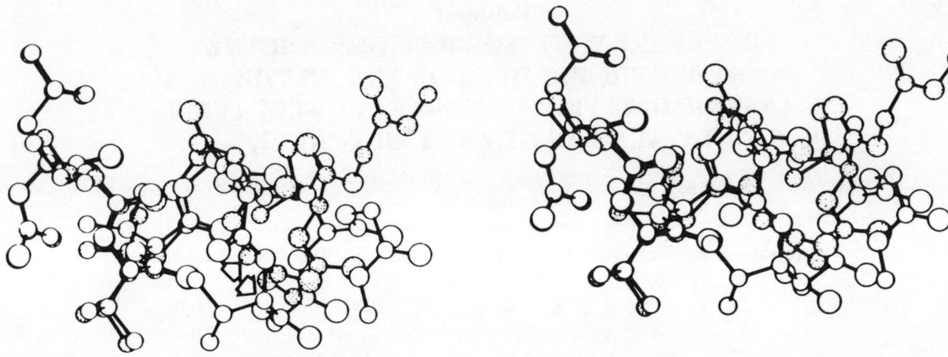

FIGURE 3. Superimposed stereo views of the computed lowest energy structures for the transforming region of the P21 protein (Leu 6-Gly 15) of the peptide from the normal protein (stippled circles) with Gly 12 and from the abnormal protein (open circles) with Val 12. The arrows point to the Cα's of residues 12 for these two peptides. The amino terminus for both structures is to the left.

12-peptide absorbs at two different field strengths, one identical to that of the Ala region of the Gly 12-peptide and another of lower weight at higher field. This second peak represents different conformer of the Val 12 peptide. It is likely that the affected Ala residue is Ala 11, the critical residue which terminates the helix in the "normal" conformation of the protein (see above). Experiments are currently being performed to determine the actual three-dimensional structure of each peptide using NH-CH coupling constants and nuclear overhauser effects (NOE) in two-dimensional NMR experiments.

1. Effects of Other Substitutions at Position 12

The above results suggest correlation between structure and function. We have therefore investigated the structural effects of other amino acid substitutions for Gly at position 12; in particular Lys, Ser, and Pro[69-73] were investigated. The results for peptides with Lys and Ser at position 12 are shown in Tables 4 and 5, respectively. It may at once be observed that the global minima for these two peptides are the same as for the Val 12-peptide despite the nonconservative nature of the substitutions. The same results (not shown here) have been obtained for the peptide with Ile, Leu, Asp, and Ala at position 12.[78] This result suggests the existence of a "malignancy-causing" structure.[69-73] As may also be noted in the two tables, both Lys- and Ser-containing peptides are more α-helical than the Gly 12-peptide (Table 2) and more constrained as is the case for the Val 12-peptide (Table 3). No structures were obtained for the Val-, Lys-, and Ser-containing peptides (Tables 3 to 5, respectively) in which the α-helix terminates at residues 11 and 12 and in which residue 12 adopts the D* conformation.

The results (Table 6) with Pro at position 12, however, are distinctly different.[69-73] As may be seen in this table, many structures such as Conformers 1 and 2, have α-helixes that terminate at residue 11 and 12. No structures occur in which the α-helix propagates through residue 11 in contrast to the results with the Val-, Lys-, and Ser-peptides. As found in the cases of the leader sequence[59] and melittin,[38] Pro causes disruption of α-helixes by forcing the residue preceding it to adopt a non-α-helical conformation. It may be further observed in Table 6 that a number of nonhelical structures occur such as Conformer 10.

Because the Pro 12-peptide does not adopt the same conformation as the peptides with the other L-amino acids (such as Conformers 1 in Tables 3 to 5), but does adopt conformations in which helix termination occurs at residues 11 and 12 (as occurs for the Gly 12-peptide) it is likely that the Pro 12-peptide has a low transformation potential. This result has been found in the experiments alluded to earlier in which each of the P21 proteins containing a

Table 4
LOW ENERGY CONFORMATIONS FOR THE HYDROPHOBIC DECAPEPTIDE IN THE TRANSFORMING PROTEINS *N*-ACETYL-LEU-(VAL)$_3$-GLY-ALA-LYS-GLY-VAL-GLY-NHCH$_3$

	Conformational state										
Conformer	**Leu**	**Val**	**Val**	**Val**	**Gly**	**Ala**	**Lys**	**Gly**	**Val**	**Gly**	**Energy (kcal/mol)**
1	A	A	A	A	A	A	C	D*	A	A	0.0
2	A	A	A	A	A	A	A	A	A	A	0.0
3	A	A	A	A	A	A	A	A	A	G*	0.1
4	A	A	A	A	A	A	A	A	C	D*	0.7
5	A	A	A	A	A	A	C	D*	A	A*	0.7
6	A	A	A	C	D*	A	A	A	A	A	0.7
7	A	A	A	A	A	A	A	A	A	C	0.7
8	A	A	A	A	A	A	A	A	A	A*	0.8
9	A	A	A	A	A	A	D	D*	A	A	1.1
10	A	A	A	A	A	A	C	D*	A	G*	1.5
11	A	A	A	A	A	A	A	A	C	C	1.6
12	A	A	A	C	D*	A	C	D*	A	A	1.8
13	A	A	A	C	D*	A	A	A	A	G*	2.0

Note: See legend to Table 1.

Table 5
LOW ENERGY CONFORMATIONS FOR THE HYDROPHOBIC DECAPEPTIDE IN THE TRANSFORMING PROTEIN *N*-ACETYL-LEU-(VAL)$_3$-GLY-ALA-SER-GLY-VAL-GLY-NHCH$_3$

	Conformation state										
Conformer	**Leu**	**Val**	**Val**	**Val**	**Gly**	**Ala**	**Ser**	**Gly**	**Val**	**Gly**	**Energy (kcal/mol)**
1	A	A	A	A	A	A	C	D*	A	A	0.0
2	A	A	A	A	A	A	A	A	C	A*	1.5
3	A	A	A	A	A	A	A	A	A	A*	1.9
4	A	A	A	A	A	A	A	A	A	C*	2.0

Note: See legend to Table 1.

different one of all of the naturally occurring amino acids at position 12 was found to induce malignant tumors except the one with Pro (and Gly) at this position.

The differences in structures of the global minima for the Lys 12- and Pro 12-peptides may be seen by comparing their space-filling stereo views in Figures 4 and 5. The results of these calculations suggest that D-amino acids at position 12 that can terminate the α-helix at positions 11 and 12 and can adopt the D* conformation may "lock" the protein into a nonmalignant form.[70,71] This prediction remains to be tested. The correlations between the structure calculations and genetic experiments are summarized in Table 7.

2. Secondary Structure Predictions

It is interesting that recent secondary structure prediction schemes also indicate that the substitution of amino acids like Val or Asp for Gly 12 result in discrete changes in secondary structures as does the substitution of Leu for Gln 61 (References 5 and 74).[79] All of the

Table 6
LOW ENERGY CONFORMATIONS FOR THE HYDROPHOBIC DECAPEPTIDE IN THE TRANSFORMING PROTEIN *N*-ACETYL-LEU-(VAL)$_3$-GLY-ALA-PRO-GLY-VAL-GLY-NHCH$_3$

	Conformational state										
Conformer	**Leu**	**Val**	**Val**	**Val**	**Gly**	**Ala**	**Pro**	**Gly**	**Val**	**Gly**	**Energy (kcal/mol)**
1	A	A	A	A	A	D	A	A	A	C	0.0
2	A	A	A	A	A	D	A	A	A	C*	0.2
3	A	A	A	A	A	D	C	D*	A	A	0.3
4	A	A	A	A	A	D	C	D*	A	C*	0.8
5	A	A	A	A	A	D	E	C*	A	A	0.4
6	A	A	A	A	A	D	A	A	A	A	1.0
7	A	A	A	A	A	D	C	D*	A	C	1.2
8	A	A	A	A	A	D	E	C*	A	C	1.3
9	A	A	A	A	A	D	A	A	A	D	1.7
10	A	A	A	A	A	D	C	D*	C	A*	2.0

Note: See legend to Table 1.

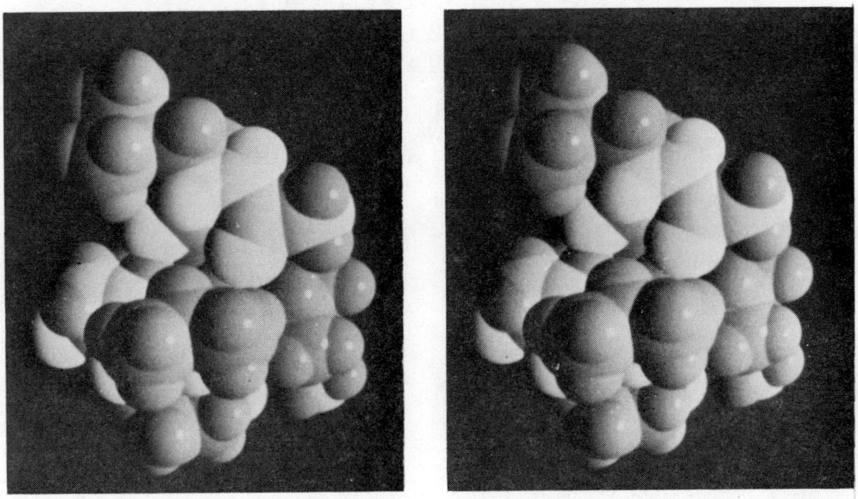

FIGURE 4. Space-filling stereo view of the P21 peptide Leu 6-Gly 15 with Lys at position 12 (amino terminus is on the lower left of the figure).

secondary structure prediction schemes indicate that the first ten residues of the P21 protein are α-helical followed by a discrete β-turn region from residue 10-13 with a peak in β-turn probability at Gly 12 (see, for example, Reference 74). Substitution of Val for Gly 12 results in extension of the α-helix through and past Gly 12 with a relatively *low* bend probability at residue 12. These results clearly indicate that major structural changes occur upon substitution of a variety of L-amino acids for Gly 12 and that α-helix propagation though residues 11 and 12 may be associated with malignant transformation, as indeed found in the computations of the three-dimensional structures of the hydrophobic decapeptide.[68-73] However, the secondary structure analysis cannot provide an explanation for the fact that high intracellular levels of the normal P21 protein with Gly at position 12 cause cell transformations.[11] The results of the conformational energy calculations are suggestive of the explanation[68-73] as described above. Nonetheless, the analysis of secondary structure of the P21 protein

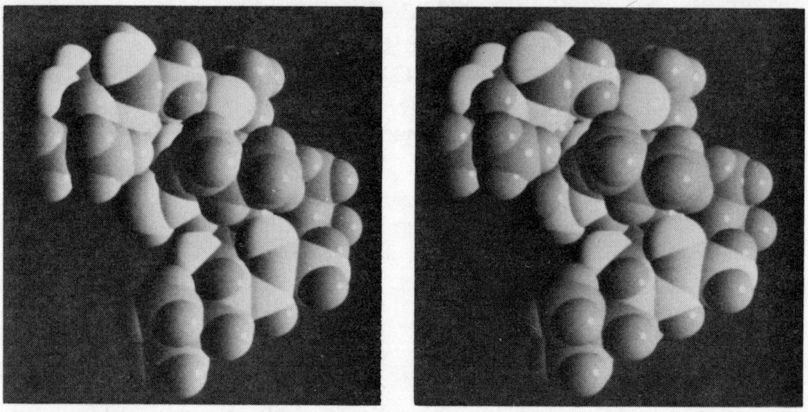

FIGURE 5. Space-filling stereo view of the P21 peptide Leu 6-Gly 15 with Pro at position 12 (amino terminus is on the lower left of the figure). This structure is considerably more open than the one in Figure 4.

Table 7
COMPARISON OF RESULTS FOR COMPUTED PREFERRED STRUCTURES FOR THE P21 PROTEIN SEQUENCE LEU 6-GLY 15 WITH EXPERIMENTAL RESULTS

Theoretical	Experimental
In the "normal" protein, α-helix ends in CD* conformation at Ala 11-Gly 12, unique in Gly 12-containing protein; L-Amino acids at position 12 cause significant structural terations	Random substitutions at position 12 of L-amino acids for Gly in P21 proteins cause malignant transformation
L-Amino acids such as Lys and Ser but not Pro substituting for Gly 12 all have the same lowest energy conformation; Pro promotes break in helix at Ala 11-Pro 12, same as for "normal" protein	All L-amino acids that substitute for Gly 12, except for Pro cause malignant transformation of cells with high efficiency
Because Gly 12-containing P21 protein can adopt "malignancy-causing" conformation with higher energy, high concentrations of "normal" P21 protein should cause transformation	Synthetic gene with viral LTR spliced to proto-oncogene causes high intracellular levels of P21 and malignant transformation

allows rapid evaluation of possible structural consequences of critical amino acid substitutions in the P21 proteins.

3. Results on Conformational Effects of Substitutions at Position 13 in the Hydrophobic Decapeptide

It has been observed that *ras* genes coding for P21 proteins that contain certain substitutions for Gly 13 (and presumably the proteins themselves) are capable of transforming NIH 3T3 cells.[6] Substitution of Ser for Gly 13 results in a protein which transforms with a slightly increased efficiency compared with that of the normal proteins but with a greatly diminished efficiency compared with that of the EJ protein (with Val for Gly 12).[6] Substitution of Asp

Table 8
LOW ENERGY CONFORMATIONS FOR THE HYDROPHOBIC DECAPEPTIDE IN THE NONTRANSFORMING PROTEIN *N*-ACETYL-LEU-(VAL)$_3$-GLY-ALA-GLY-SER-VAL-GLY-NHCH$_3$

	Conformational state										
Conformer	**Leu**	**Val**	**Val**	**Val**	**Gly**	**Ala**	**Gly**	**Ser**	**Val**	**Gly**	**Energy (kcal/mol)**
1	A	A	A	A	A	C	D*	A	A	C	0.0
2	A	A	A	A	A	A	A	A	A	A	0.3
3	A	A	A	A	A	A	A	A	A	C	0.4
4	A	A	A	A	A	A	A	A	A	A*	0.6
5	A	A	A	A	A	A	A	A	C	D*	0.7
6	A	A	A	A	A	A	A	A	A	G*	0.7
7	A	A	A	A	A	A	A	A	A	D*	1.3
8	A	A	A	A	A	A	A	A	A	C*	1.4
9	A	A	A	A	A	C	D*	A	A	A*	1.4
10	A	A	A	A	A	C	D*	A	A	D*	1.5
11	A	A	A	A	A	A	A	A	C	A*	1.8
12	A	A	A	A	A	C	D*	A	A	A	1.9
13	A	A	A	A	A	C	D*	A	A	A*	2.0

Note: See legend to Table 1.

Table 9
LOW ENERGY CONFORMATIONS FOR THE HYDROPHOBIC DECAPEPTIDE IN THE TRANSFORMING PROTEIN *N*-ACETYL-LEU-(VAL)$_3$-GLY-ALA-GLY-ASP-VAL-GLY-NHCH$_3$

	Conformational state										
Conformer	**Leu**	**Val**	**Val**	**Val**	**Gly**	**Ala**	**Gly**	**Asp**	**Val**	**Gly**	**Energy (kcal/mol)**
1	A	A	A	A	A	C	G	A	A	C	0.0
2	A	A	A	A	A	C	G	A	A	A*	1.2
3	A	A	A	A	A	A	A	A	A	A	1.2
4	A	A	A	A	A	A	A	A	A	G*	1.4
5	A	A	A	A	A	C	G	A	A	C*	1.8
6	A	A	A	A	A	C	G	A	A	A	1.8

Note: See legend to Table 1.

for Gly 13 results in a protein which transforms with notably higher efficiency than the normal protein but with significantly lower efficiency than that of the EJ protein.[6] Recent calculations on the preferred structures for the hydrophobic decapeptide (Leu 6-Gly 15) indicate that the Ser 13-containing peptide adopts as its global minimum the "normal" peptide conformation (Conformer 1 in Table 8) as found for the normal protein (Conformer 1, Table 2), but that alternate conformations now tentatively associated with cell transformation, in which the helix propagates through Ala 11, occur (e.g., Conformer 2 in Table 8). Results for the decapeptide with Asp 13, as shown in Table 9, indicate that the global minimum is a structure in which Gly 12 adopts the G rather than D* conformation. This conformation involves a change in the dihedral angle ϕ of Gly 12 from 170° to $-160°$, a change in 30°. Attempts to reset the Gly 12 dihedral angles for the D* state result in a high energy conformation which returns to the G conformation under energy minimization. The high energy for the Asp 13-peptide with Gly 12 in the D* conformation is caused by bad contacts between the side chain atoms of the –COOH group of Asp 13 and backbone atoms of Val 9, which can be relieved best by the D*→ G structure change for Gly 12. Thus, the Asp 13-containing peptide cannot adopt the same lowest energy structure as the Gly 12-containing peptide. In Table 9, it may be seen that it can adopt conformations such as Conformer 2 in which the helix propagates through residue 11. These results suggest that amino acids with highly branched side chains such as Val and Ile will likewise induce similar changes in structure as those caused by Asp 13. Therefore, proteins containing such substitutions may exhibit increased transforming capacities. Computed results with Val at position 13 show that the distribution of low energy minima is the same as with Asp at this position and the global minima are the same.[78]

Thus, the P21 protein-containing substitution of Val for Gly 13 would be expected to cause significant transformation. Experimentally, the *ras*-genes from patients with acute myelogenous leukemia have been cloned and sequenced.[75] It has been found that the base sequence is identical to that of the proto-oncogene except that Val substitutes for Gly at position 13. This cloned gene has been transfected into NIH 3T3 cells and causes transformation.[75] This finding appears to lend support to the conclusions from conformational analysis.

4. Conclusions on the Structure and Activity of the Hydrophobic Decapeptide Leu 6-Gly 15

The correlations obtained between structure and activity for the peptides with Gly and other amino acids substituted at position 12 are summarized in Table 7. These correlations strongly suggest that there are two basic structures that can be adopted by this segment: the normal form with a kink in the chain at Ala 11 and a "malignancy-causing" form with helix propogation through Ala 11. Effects of amino acid substitutions for Gly at position 13 result either in preservation of the "normal" conformation (Conformer 1, Table 2) or in changes in conformation at Gly 12 preventing adoption of the "normal" conformation and allowing expression of "malignancy causing" structures. A simple scheme summarizing these conclusions is presented in Figure 6. In this figure, the Gly-12-containing protein is shown to equilibrate between a normal and a transformating conformation. In the normal conformation, the protein binds to a regulatory "cell element" E, (which may in fact be more than one molecule) with high affinity. With sufficient E present, all normal P21 protein is bound to E in the normal conformation. In Figure 6, this complex is represented as Glycine 12-E. The complex would also exist for Pro 12-E. With proteins containing noncyclic L-amino acids at position 12, the "normal" E-binding conformation is no longer present; the protein no longer binds E, and cell transformation results represented in Figure 6 as Valine 12 and X-12 (other noncyclic L-amino acids at position 12) in the lower part of the figure. In the event that E is decreased in concentration *or* the normal protein is present in abnormally elevated concentration, saturation of E by normal protein ensues, and a significant number of free P21 molecules exists. Equilibration between normal and abnormal forms results with either or both forms free to cause cell transformation. Thus, transformation can be effected by any or all of a number of different paths such as presence of abnormal protein, high intracellular levels of normal protein or abnormally low concentrations, or abnormal non-binding forms of "E".

B. Structures of the Central Transforming Region Containing Residues 59, 61, and 63

Computations on the structure of a 13-residue segment containing three critical residues are currently being performed. This segment is Ile 55-Met 67 whose sequence is

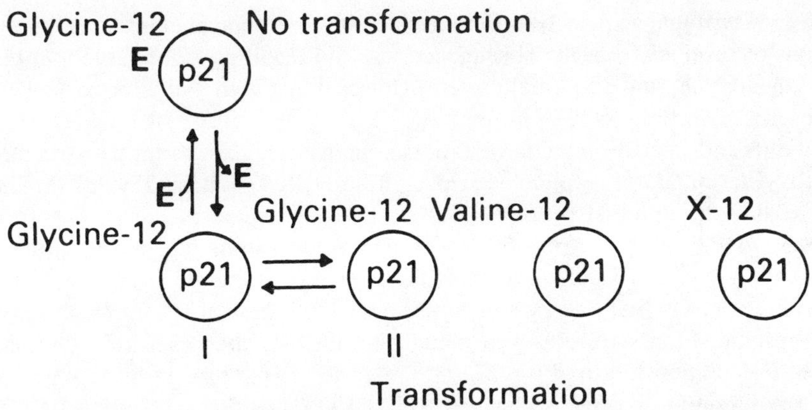

FIGURE 6. Proposed mechanism by which the P21 protein may cause cell transformation. Two conformations, I (normal) and II (abnormal), exist for the normal protein. At usual intracellular concentrations of the normal protein, the normal structure allows for binding to a cell element E such that *all* P21 molecules are so bound (no transformation). Elevation of normal protein concentration intracellularly results in saturation of E and free P21 molecules that can exist in the "malignancy-causing conformations" and cause cell transformation ("Transformation" in figure). The abnormal protein (like the Val 12-containing protein) cannot adopt the normal conformation and hence cannot bind E. X-12 refers to P21 proteins with all L-amino acids, except Pro at position 12, none of which can adopt structures capable of binding E and hence which cause cell transformation.

```
                        59        61        63
Ile-Leu-Asp-Thr-Ala-Gly-Gln-Glu-Glu-Tyr-
                Thr     Lys     Lys
                        Leu
Ser-Ala-Met
```

The three critical residues are underlined and are labeled. Malignancy-causing substitutions are listed beneath each one. This particular segment was selected as a starting sequence for conformational analysis because it appears to be a major antigenic determinant.[21,22] In addition, studies have been performed on the minimal length segment of neighboring amino sequences on either side (amino and carboxyl) of a central residue that determines the conformational state of that central residue.[76] The peptide 55-67 was chosen therefore also because each of the critical residues (59, 61, and 63) is flanked by a minimum number of four neighboring residues on both amino and carboxyl terminal ends. In the absence of solvent, we have found that the unsubstituted sequence is highly structured, residues 55-61 adopting a β-conformation while residues 62-67 adopt an α-helical structure.[80] With substitution of Lys for Gln at position 61, the peptide is found to undergo a structural change in which the helix propogates from residue 60-67 while β-structure is confined to residues 55-59. Most interestingly, the same result has been obtained when Leu substitutes for Gln 61, again suggesting a discrete structural effect on this peptide segment, in which structure correlates with transforming activity. Similar conclusions have been reached through the use of secondary structure analysis.[74]

C. Structures of the Carboxyl Terminal Decapeptides of the P21 Protein

As discussed in the Introduction (Section I.B.3), strong experimental evidence exists that the carboxyl terminal decapeptides for a variety of different P21 proteins such as Harvey (H), Kirsten (KA and KB), and N-*ras* proteins are critical in membrane binding. Though these carboxyl terminal sequences differ considerably from one another, substitution in and/or deletions of residues from the H sequence result in inability of the protein to bind to the inner cell membrane, and abnormal proteins (for example with Val at position 12) are then incapable of transforming cells.

We have recently performed conformational energy calculations on the structures of the four carboxyl terminal decapeptide sequences from H, KA, KB, and N-*ras* proteins.[80] We have found that the terminal pentapeptides for all four sequences are all α-helical. In the case of KA-*ras* P21 protein, from 5000 starting conformations, only three structures have energies that lay within a 5 kcal/mol cutoff of the energy of the global minimum which was an α-helix. The two higher energy structures were also α-helical except for the first (amino terminal) residue. Similar results were obtained with the other three (H, KB, and N-*ras*) sequences. Calculations on the PBW277 mutant of the H-*ras* protein in which the carboxyl terminal Lys-Cys-Val-Leu-Ser is changed to Lys-Cys-Thr-Pro (see Section I.B.3) demonstrate that the helix is broken at Thr in the PBW277 mutant. This result is expected because of the role of Pro in helix disruption as found in the cases of leader sequence, melittin, and the P21 proteins discussed previously. As noted in the Introduction, the PBW277 mutant does not bind to the cell membrane and does not transform cells[17,18] indicating that for the last five residues at least, an α-helix is the critical structure. Thus far, there seems to be a strong structure-function correlation for these peptides also.

D. Conclusion

From the above-described results, a strong pattern emerges from each of the three distinct domains of the P21 protein implicated in transforming activity, residues 6-15, 55-67, and 180-189. Characteristic structural changes are induced by amino acid substitutions that cause malignant transformation of cells in the first two sequences. Amino acid substitutions that disrupt the carboxyl terminal α-helix in the P21 protein consequently prevent membrane-binding of the P21 protein with loss of transforming activity. The results of these calculations are encouraging and will now enable extension of these methods of analysis to predict possible other regions of the protein that may be implicated in cell transformation. In addition, the buildup method can be readily extended to computation of larger peptide segments of this protein and eventually can be applied to the entire P21 sequence.

REFERENCES

1. **Tabin, C. J., Bradley, S. M., Bargmann, C. I., Weinberg, R. A., Papageorge, A. G., Scolnick, E. M., Dhar, R., Lowy, D. R., and Chang, E. H.,** Mechanism of activation of a human oncogene, *Nature (London),* 300, 143, 1982.
2. **Reddy, E. P., Reynolds, R. K., Santos, E., and Barbacid, M.,** A point mutation is responsible for the acquisition of transforming properties by the T24 human bladder carcinoma oncogene, *Nature (London),* 300, 149, 1982.
3. **Capon, D. J., Chen, E. Y., Levinson, A. D., Seeburg, P. H., and Goeddel, D. V.,** Complete nucleotide sequences of the T24 human bladder carcinoma oncogene and its normal homologue, *Nature (London),* 302, 33, 1983.
4. **Stacy, D. W. and Kung, H. F.,** Transformation of NIH 3T3 cells by microinjection of Ha-ras P21 protein, *Nature (London),* 310, 508, 1984.
5. **Santos, E., Reddy, E. P., Pulcianai, S., Feldmann, R. J., and Barbacid, M.,** Spontaneous activation of a human proto-oncogene, *Proc. Natl. Acad. Sci. U.S.A.,* 80, 4679, 1983.
6. **Fasano, O., Aldrich, T., Tamanoi, F., Taparowsky, E., Furth, M., and Wigler, M.,** Analysis of transforming potential of the human H-ras gene by random mutagenesis, *Proc. Natl. Acad. Sci. U.S.A.,* 81, 4008, 1984.
7. **McGrath, J. P., Capon, D. J., Goeddel, D. V., and Levinson, A. D.,** Comparative biochemical properties of normal and activated human *ras* P21 protein, *Nature (London),* 644, 1984.

8. **Gibbs, J. B., Sigal, I. S., Poe, M., and Scolnick, E. M.,** Intrinsic GTPase activity distinguishes normal and oncogenic *ras* P21 molecules, *Proc. Natl. Acad. Sci. U.S.A.*, 81, 5704, 1984.
9. **Hurley, J. B., Simon, M. I., Teplow, D. B., Rabishaw, J. D., and Gilman, A. G.,** Homologies between signal transducing G protein and *ras* gene products, *Science*, 226, 860, 1984.
10. **Kataoka, T., Powers, S., Cameron, S., Fasano, O., Goldfarb, M., Broach, J., and Wigler, M.,** Functional homology of mammalian and yeast *ras* genes, *Cell*, 40, 19, 1985.
11. **Chang, E. H., Furth, M. E., Scolnick, E. M., and Lowy, D. R.,** Tumorigenic transformation of mammalian cells induced by a normal human gene homologous to the oncogene of Harvey murine sarcoma virus, *Nature (London)*, 297, 479, 1982.
11a. **Kung, H. F.,** personal communication.
12. **Gay, N. J. and Walker, J. E.,** Homology between human bladder carcinoma oncogene product and mitochondrial ATP-synthetase, *Nature (London)*, 301, 260, 1983.
13. **Wierenga, R. K. and Hol, W. G. J.,** Predicted nucleotide binding properties of P21 protein and its cancer-associated variant, *Nature (London)*, 302, 842, 1983.
14. **Seeburg, P. H., Colby, W. W., Capon, D. J., Goedddel, D. N., and Levinson, A. D.,** Biological properties of human c-Ha-*ras* 1 genes mutated at codon 12, *Nature (London)*, 71, 312, 1984.
15. **Taparowsky, E., Shimizu, K., Goldfarb, M., and Wigler, M.,** Structure and activation of the human N-ras gene, *Cell*, 34, 581, 1983.
16. **Leberman, R. and Egner, V.,** Homologies in the primary structure of GTP-binding proteins: the nucleotide-binding site of EF-tu and P21, *EMBO J.*, 3, 339, 1984.
17. **Willumsen, B. M., Christensen, A., Hubbert, N. L., Papageorge, A. G., and Lowy, D. R.,** The P21 *ras*-C-terminus is required for transformation and membrane-association, *Nature (London)*, 310, 583, 1984.
18. **Willumsen, B. M., Norris, K., Papageorge, A. G., Hubbert, N. L., and Lowy, D. R.,** Harvey murine sarcoma virus P21 *ras* protein: biological and biochemical significance of the cysteine residue nearest the carboxy terminus, *EMBO J.*, 3, 2581, 1984.
19. **Bolanowski, M., Earles, B. J., and Lennarz, W. J.,** Fatty acylation of protein during development of sea urchin embryos, *J. Biol. Chem.*, 259, 4934, 1984.
20. **Willumsen, B. and Lowe, D. R.,** P21 ras transforming protein: significance of the carboxy terminus, *Dev. Oncol.*, 28, 25, 1985.
21. **Furth, M. E., Davis, L. J., Fleurdelys, B., and Scolnick, E. M.,** Monoclonal antibodies to the P21 products of the transforming gene of Harvey murine sarcoma virus and of the cellular *ras* gene family, *J. Virol.*, 43, 294, 1982.
22. **Feramisco, J. R., Clark, R., Wong, G., Arnheim, N., Milley, R., and McCormick, F.,** Transient reversion of *ras* oncogene-induced cell transformation by antibodies specific for amino acid 12 of *ras* protein, *Nature (London)*, 314, 639, 1985.
23. **Anfinsen, C. B. and Scheraga, H. A.,** Experimental and theoretical aspects of protein folding, *Adv. Protein Chem.*, 29, 205, 1975.
24. **Némethy, G. and Scheraga, H. A.,** Protein folding, *Q. Rev. Biophys.*, 10, 239, 1977.
25. **Scheraga, H. A.,** Protein structure and function, from a colloidal to a molecular view, *Carlsberg Res. Commun.*, 49, 1, 1984.
26. **Scheraga, H. A.,** Recent progress in the theoretical treatment of protein folding, *Biopolymers*, 22, 1, 1983.
27. **Scheraga, H. A.,** Interactions in polypeptides and proteins, *Jerusalem Symp. Quantum Chem. Biochem.*, 5, 51, 1973.
28. **Momany, F. A., Carruthers, L. M., McGuire, R. F., and Scheraga, H. A.,** Intermolecular potentials from crystal data. III. Determination of empirical potentials and application to the packing configurations and lattice energies in crystals of hydrocarbons, carboxylic acids, amines, and amides, *J. Phys. Chem.*, 78, 1595, 1974.
29. **Momany, F. A., McGuire, R. F., Burgess, A. W., and Scheraga, H. A.,** Energy parameters in polypeptides. VII. Geometric parameters, partial atomic charges, non-bonded interactions, hydrogen bond interactions, and intrinsic torsional potentials for the naturally occurring amino acids, *J. Phys. Chem.*, 79, 2361, 1975.
30. **Némethy, G., Pottle, M., and Scheraga, H. A.,** Energy parameters in polypeptides. IX. Updating of geometrical parameters, non-bonded interactions, and hydrogen bond interactions for the naturally occurring amino acids, *J. Phys. Chem.*, 87, 1883, 1983.
31. **Pincus, M. R. and Scheraga, H. A.,** An approximate treatment of long range interactions in protein, *J. Phys. Chem.*, 81, 1579, 1977.
32. **Hodes, Z. I., Némethy, G., and Scheraga, H. A.,** Model for the conformational analysis of hydrated peptides. Effect of hydration on the conformational stability of the terminally blocked residues of the 20 naturally occurring amino acids, *Biopolymers*, 18, 1565, 1979.
33. **Hodes, Z. I., Némethy, G., and Scheraga, H. A.,** Influence of hydration on the conformational stability and formation of bends in terminally blocked dipeptides, *Biopolymers*, 18, 1611, 1979.

34. **Zimmerman, S. S., Pottle, M. S., Némethy, G., and Scheraga, H. A.,** Conformational analysis of the 20 naturally occurring amino acid residues using ECEPP, *Macromolecules,* 10, 1, 1977.
35. **Vásquez, M., Némethy, G., and Scheraga, H. A.,** Computed conformational states of the 20 amino acid residues and of the prototype residue α-aminobutyric acid, *Macromolecules,* 16, 1043, 1983.
36. **Pincus, M. R., Burgess, A. W., and Scheraga, H. A.,** Conformational energy calculations of enzyme-substrate complexes of lysozyme. I. Energy minimization of mono- and oligosaccharide inhibitors and substrates of lysozyme, *Biopolymers,* 15, 2485, 1976.
37. **Powell, M. J. D.,** An efficient method for finding the minima of a function of several variables without calculating derivatives, *Comput. J.,* 7, 155, 1964.
38. **Pincus, M. R., Klausner, R. D., and Scheraga, H. A.,** Calculation of the three-dimensional structure of the membrane-bound portion of melittin from its amino acid sequence, *Proc. Natl. Acad. Sci. U.S.A.,* 79, 5707, 1982.
39. **Dygert, M., Go, N., and Scheraga, H. A.,** Use of a symmetry condition to compute the conformation of gramicidin S, *Macromolecules,* 8, 750, 1975.
40. **Miller, M. H. and Scheraga, H. A.,** Calculation of the structure of collagen models. Role of interchain interactions in determining the triple-helical coiled-coil conformation. I. Poly (Glycyl-Prolyl-Prolyl), *J. Polym. Sci. Polym. Symp.,* 54, 171, 1976.
41. **Hull, S. E., Karlsson, R., Main, P., Woolfson, M. M., and Dodson, E. J.,** The crystal structure of a hydrated gramicidin S-urea complex, *Nature (London),* 275, 206, 1978.
42. **Naik, V. M., Krimm, S., Denton, J. B., Némethy, G., and Scheraga, H. A.,** Vibrational analysis of peptides, polypeptides, and proteins, *Int. J. Pept. Protein Res.,* 24, 613, 1984.
43. **Traub, W. and Piez, K. A.,** The chemistry and structure of collagen, *Adv. Prot. Chem.,* 25, 243, 1971.
44. **Okuyama, K., Tanaka, N., Ashida, T., and Kakudo, M.,** Structure analysis of a collagen model polypeptide, $(\text{Pro-Pro-Gly})_{10}$, *Bull. Chem. Soc. Jpn.,* 49, 1805, 1976.
45. **Miller, M. H., Némethy, G., and Scheraga, H. A.,** Calculation of the structures of collagen models. Role of interchain interactions in determining the triple helical coiled-coil conformation. III. Poly (Glycyl-prolyl-alanyl), *Macromolecules,* 13, 910, 1980.
46. **Némethy, G., Miller, M. H., and Scheraga, H. A.,** Calculation of the structures of collagen models. Role of interchain interactions in determining the triple helical coiled-coil conformation. IV. Poly (Glycyl-alanyl-prolyl), *Macromolecules,* 13, 914, 1980.
47. **Pincus, M. R. and Scheraga, H. A.,** Theoretical calculations on enzyme-substrate complexes: the basis of molecular recognition and catalysis, *Acc. Chem. Res.,* 14, 299, 1981.
48. **Scheraga, H. A., Pincus, M. R., and Burke (Platzer), K. E.,** in *Structure of Complexes Between Biopolymers and Low Molecular Weight Molecules,* Bartmann, W. and Snatzke, G., Eds., John Wiley & Sons, New York, 1982, 3.
49. **Platzer, K. E. B., Momany, F. A., and Scheraga, H. A.,** Conformational energy calculations of enzyme-substrate complexes. II. Computation of the binding energy for substrates in the active site of α-chymotrypsin, *Int. J. Pept. Protein Res.,* 4, 201, 1972.
50. **Pincus, M. R. and Scheraga, H. A.,** Conformational energy calculations of enzyme substrate and enzyme-inhibitor complexes of lysozyme. II. Calculation of the structures of complexes with a flexible enzyme, *Macromolecules,* 12, 633, 1979.
51. **Pincus, M. R. and Scheraga, H. A.,** Prediction of the three-dimensional structures of complexes of lysozyme with cell wall substrates, *Biochemistry,* 20, 3960, 1981.
52. **Pincus, M. R., Zimmerman, S. S., and Scheraga, H. A.,** Prediction of three dimensional structures of enzyme-substrate and enzyme-inhibitor complexes of lysozyme, *Proc. Natl. Acad. Sci. U.S.A.,* 73, 4261, 1976.
53. **Pincus, M. R., Zimmerman, S. S., and Scheraga, H. A.,** Structures of enzyme-substrate complexes of lysozyme, *Proc. Natl. Acad. Sci. U.S.A.,* 74, 2629, 1977.
54. **Imoto, T., Johnson, L. N., North, A. C. T., Phillips, D. C., and Rupley, J. A.,** Vertebrate lysozymes, in *The Enzymes,* Vol. 7, 3rd. ed., Boyer, P. D., Ed., Academic Press, New York, 1972, 665.
55. **Schindler, M., Assaf, Y., Sharon, N., and Chipman, D. M.,** Mechanism of lysozyme catalysis: role of ground-state strain in subsite D in hen egg-white and human lysozymes, *Biochemistry,* 16, 423, 1977.
56. **Kelly, J. A., Sielecki, A. R., Sykes, B. D., and James, M. N. G.,** X-ray crystallography of the binding of the bacterial cell wall trisaccharide NAM-NAG-NAM to lysozyme, *Nature (London),* 282, 875, 1979.
57. **Smith-Gill, S. J., Rupley, J. A., Pincus, M. R., Carty, R. P., and Scheraga, H. A.,** Experimental identification of a theoretically-predicted "left-sided" binding mode for $(\text{GlcNAc})_6$ in the active site of lysozyme, *Biochemistry,* 23, 993, 1984.
58. **Smith-Gill, S. J., Wilson, A. C., Potter, M., Feldmann, R. J., and Mainhart, C. R.,** Mapping the antigenic epitope for a monoclonal antibody against avian lysozyme, *J. Immunol.,* 128, 314, 1982.
59. **Pincus, M. R. and Klausner, R. D.,** Prediction of the three dimensional structure of the leader sequence of pre-κ light chain, a hexadecapeptide, *Proc. Natl. Acad. Sci. U.S.A.,* 79, 3413, 1982.

60. **Zimmerman, S. S. and Scheraga, H. A.,** Influence of local interactions on protein structure. I. Conformational energy studies of N-acetyl-N′-methyl amides of Pro-X and X-Pro dipeptides, *Biopolymers,* 16, 811, 1977.
61. **Rosenblatt, M., Beaudette, M. V., and Fasman, G. D.,** Conformational studies of the synthetic precursor-specific region of preproparathyroid hormone, *Proc. Natl. Acad. Sci. U.S.A.,* 77, 3983, 1980.
62. **Hortin, G. and Boime, I.,** Inhibition of preprotein processing in ascites tumor lysates by incorporation of a leucine analog, *Proc. Natl. Acad. Sci. U.S.A.,* 77, 1356, 1980.
63. **Emr, S. D. and Silhavy, T. J.,** The molecular components of the signal sequence that function in the initiation of protein export, *J. Cell Biol.,* 95, 689, 1982.
64. **Engelman, D. M. and Steitz, T. A.,** The spontaneous insertion of proteins into and across membranes: the helical hairpin hypothesis, *Cell,* 23, 411, 1981.
65. **Scheele, G.,** Biosynthesis, segregation, and secretion of exportable proteins by the exocrine pancreas, *Am. J. Physiol.,* 238, G467, 1980.
66. **Brown, L. R., Braun, W., Kumar, A., and Wüthrich, K.,** High resolution nuclear magnetic resonance studies of the conformation and orientation of melittin bound to a lipid-water interface, *Biophys. J.,* 37, 319, 1982.
67. **Terwiliger, T. C., Weissman, L., and Eisenberg, D.,** The structure of melittin in the form I crystals and its implications for melittin's lytic and surface activities, *Biophys. J.,* 37, 353, 1982.
68. **Pincus, M. R., van Renswoude, J., Harford, J. B., Chang, E. H., Carty, R. P., and Klausner, R. D.,** Prediction of the three-dimensional structure of the transforming region of the EJ/T24 human bladder oncogene product and its normal cellular homologue, *Proc. Natl. Acad. Sci. U.S.A.,* 80, 5253, 1983.
69. **Pincus, M. R.,** Protein structure and oncogenesis, in *The Molecular Basis of Cancer: An Interdisciplinary Discussion on Basic and Applied Aspects of Cancer,* Rein, R., Ed., Alan R. Liss, New York, 1985, 419.
70. **Pincus, M. R. and Brandt-Rauf, P. W.,** Correlation of the structure of the P21 proteins with their transforming potentials, *18th Middle Atl. Reg. Meet. Am. Chem. Soc. Proc.,* Am. Chem. Soc., Washington, D.C., 1984, 40.
71. **Pincus, M. R. and Brandt-Rauf, P. W.,** Structural effects of substitutions on the P21 proteins, *Proc. Natl. Acad. Sci. U.S.A.,* 82, 3596, 1985.
72. **Pincus, M. R. and Scheraga, H. A.,** Conformational analysis of biologically active polypeptides, with application to oncogenesis, *Acc. Chem. Res.,* in press.
73. **Pincus, M. R.,** Protein structure and cancer, *Cancer Investigation,* in press.
74. **Murakami, M.,** Mutations affecting the 12th and 61st amino acids of P21 protein result in decreased probability of a β-turn occurrence around the mutation positions: a prediction, *J. Theor. Biol.,* 114, 193, 1985.
75. **Bos, J. L., Tokoz, D., Marshall, C. J., Verlann-de Vries, Veeneman, G. H., van der Eb, A. J., van Boom, J. H., Janssen, J. W. G., and Steenvoorden, A. C. M.,** Amino acid substitutions at codon 13 of the N-ras oncogene in human acute myeloid leukemia, *Nature (London),* 315, 726, 1985.
76. **Ponnuswamy, P. K., Warme, P. K., and Scheraga, H. A.,** Role of medium-range interactions in proteins, *Proc. Natl. Acad. Sci. U.S.A.,* 70, 830, 1973.
77. **Feng-Ni, Kung, H. F., Pincus, M. R., and Scheraga, H. A.,** manuscript in preparation.
78. **Pincus, M. R. and Brandt-Rauf, P.,** manuscript submitted.
79. **Silverman, D.,** personal communication.
80. **Gibson, K., Pincus, M. R., Carty, R. P., Lubowsky, J., Avitable, M., and Scheraga, H. A.,** manuscript in preparation.

Chapter 3

CONFORMATIONAL CALCULATIONS ON ONCOGENE PRODUCTS

J. Garnier and B. Robson

TABLE OF CONTENTS

I. INTRODUCTION

Ever since the recent discovery that the P28^{sis} product of the v-*sis** oncogene of the Simian sarcoma virus is homologous to platelet-derived growth factor,[1,2] it has become increasingly clear that many if not all oncogenes are derived from cell systems regulating cell growth and development. In particular, the v-*erbB* oncogene of avian retroviruses (avian erythroblastosis virus), responsible for erythroblast and fibroblast transformation, is a modification of the gene encoding the receptor for epidermal growth factor.[3,4] The modifications lead to deletion of an N-terminal extracellular component of the epidermal growth factor receptor (EGFR), which presumably contains the epidermal growth factor binding site, leaving a truncated extracellular segment with 2 of the original 12 potential glycosylation sites. The hydrophobic transmembrane segment persists, but the intracellular component with protein kinase homology is truncated at the C-terminus, causing loss of one of five autophosphorylation sites. It appears that these changes imply a "permanently switched-on" hormone receptor, in contrast to the P28^{sis} product of the v-*sis* oncogene which acts as a hormone. In any event, v-*erbB* induces cell renewal without blocking the cell's capacity to differentiate and appears to remove the dependence on endogenous host hormones such as erythropoietin, and evidently acts by insertion in the membrane in the manner of the natural receptor.[5,6]

The overall process of cancer cell development is complex, though not obviously promising greater complexity than other cellular processes that are more or less well understood. The state of the cell is important: a particular transformation is not always possible beyond a certain level of cell maturation. One oncogene alone might be insufficient to span the full range of component mechanisms leading to full development of transformed cells. For example, virus MH2 appears to have a v-*myc* gene responsible for macrophage transformation, *and* a v-*mil* gene releasing the dependence on endogenous host growth factors. In contrast is the case when two or more properties are inherent in the *same* oncogene. The FBR-MuSV fos oncogene product exhibits properties both of transforming and "immortalizing" genes.

The latter aspect is of particular interest and highlights the need for a deeper understanding of the product molecular structure and of the component parts with their apparently distinct functions. In this respect there is little doubt that the multiple functions of an oncogene may sometimes reside in distinct functional regions of the protein product. The transforming ability of the v-*erbB* product seems to reside in, or depend on, the presence of a 129 amino acid fragment from C-199 to C-70 (i.e., the 70th to the 199th residue from the C-terminus of EGF receptor). The C-189 to C-144 fragment appears to be concerned with fibroblast transformation, and C-144 to C-70 with erythroblast transformation.[7]

A more immediate action in oncopharmaceutics may profitably concentrate on extracellular proteins (e.g., P28^{sis}) or extracellular domains (e.g., v-*erbB* oncogene product domain I), and involve the development of artificial vaccines giving selective immunity against the oncogene product over its endogenous host counterpart. While the merits of inhibitors including tumor inhibitor agents of the phorbol class should not be forgotten, these run the risk, at least in general, of interfering with natural endogenous cell growth and development mechanisms.

Whatever the route taken in the pursuit of prevention or cure in cancer biology, or in the use of the cancer state as a basic research tool in cell biology, a deeper understanding of the molecular structures of oncogene products is evidently of profound value. The present review concentrates on this aspect, with some emphasis on our own status in this area.

* v- Refers to the viral gene or oncogene as opposed to the normal cellular gene or proto-oncogene which bears the prefix c-. The name of the gene, *sis,* is for *Si*mian *S*arcoma and P28 denotes the gene product of a molecular weight of 28 kdaltons.

II. PRIOR TO ENERGY CALCULATION: A HOMOLOGY SCAN

A homology scan is essentially a procedure by which information about a protein of interest can be obtained from a database of known protein sequences or DNA sequences from which the amino acid sequence of a protein product is inferred. It is in essence a scan for sequences which are similar to that of interest and may reveal a near identity in that many residues correspond exactly, or a more subtle relationship in which few if any residues correspond exactly, but where there is evidence of their coding an equivalent conformational motif. This is particularly valuable when the corresponding sequence is of known conformation, but in principle gross conformational arrangements can be inferred even without the benefit of detailed conformational data. There is an implicit underlying idea that homologous sequences are related by evolution, which is to say they evolved from divergence of ancestral genetic material. As an assumption this is not always required, but the concept can play a role in the detailed method of choice for scanning for homology, or in interpretation of the results.

Originally, attention was directed to recognition of whole protein sequences as homologous, in the sense that myoglobins are homologous to hemoglobin subunits, and α-lactalbumin is homologous to lysozyme. Two such proteins may have many residues in identical positions, when alignment is taken into account (see below). Such homologous proteins are said to belong to the same protein family, with a common ancestral origin. Within a family, however, it is frequently found that there are very few amino acid residues which are absolutely conserved and are a characteristic feature of the family. Rather, there are certain sections of sequences which change relatively little, and which are typically characteristic of the species from which the proteins are taken. Thus, cytochrome c is said to contain "taxonomic sequences" showing that the origin is a bird, an insect, or a fungus, for example. A consequence of this is that two proteins drawn from quite different species, say, proteins A and B, may not appear to be significantly homologous. In this respect it is noteworthy that two quite unrelated sequences can have something like 15% of residues in identical positions by chance, so on the basis of looking for identical residue matches an identity of some 30% residues is required to be deemed significant. Both proteins A and B may, however, have homology with at least a third protein, say protein C. Thereby, A and B can be said to show *indirect homology*. The evolutionary conclusion is that C is more closely related to the common ancestral protein, and it is this kind of reasoning which allows the ordering of the sequence data into a tree format which is identified as the evolutionary tree of the species from which the proteins are drawn. A major value of scans for homologies of whole proteins is in the construction of such trees, which in themselves have potential taxonomic applications though criticisms have been made in that multiple solutions for the best tree can be made on the basis of given similarity data. Examples of protein sequences which are more or less homologous throughout their length, though taking account of the notion of indirect homology to demonstrate the point, are ovalbumin, antithrombin III, α-1-antitrypsin, barley protein Z, and angiotensin. In this particular example, there is no outstandingly obvious relation between the functions of the proteins concerned, in stark contrast to the members of the hemoglobin-myoglobin family, and the cytochrome c family, much used in the construction of evolutionary trees. Presumably there is a common origin but substantial diversification of function. An intermediate case is between hen egg white lysozyme and α-lactalbumin, where a functional relationship is revealed by deeper consideration of the biological including histological aspects. Such families which are internally variant in function are less useful for the construction of evolutionary trees because the different functions imply different selective pressures in evolution, and hence different rates of amino acid substitution such that differences in degree of similarity are not simply related to the length of time of divergence from a common ancestor. Rather, the importance is the general concept of a common origin, and the genuine hope that there are conformational as well as sequence homologies so that by consideration of all the homologous proteins further conformational detail can be inferred. This of course will be related to the function.

Such scans and their graphical representations,[8] and indeed variations on this theme discussed below, have a sound reputation in the cancer field. The classic example is the identification of homology between the $P28^{sis}$ protein product of the simian sarcoma virus and human platelet derived growth factor.[1,2] Strictly, the viral protein is somewhat longer but the degree of homology is extensive. The notion of a common evolutionary origin is exciting because with other data it leads to the notion that transforming retroviruses have arisen in such way that they acquire the ability to induce neoplastic transformation by a limited class of mechanisms. In this case, the mechanism is the incorporation of a gene of which the product is the host hormone concerned with an active cell state, rather than the passive one characteristic of a normal cell in mature tissue.

Recently, increased attention has been given to the homologies between fragments or sections of sequence of a protein of interest, and fragments or sections of whole proteins in the database. However in some cases, the division into fragments may be somewhat artificial, and the division being just a device for fully exploiting the notion of indirect homology. For example, a sequence ab with two consecutive sections a, b is a useful concept when a is significantly homologous to fragment c in cd, b is significantly homologous to f in ef, and sequences cd and ef are just significantly homologous. The case for the three being homologous is strengthened. In other cases, the fragments have a more clear real identity. One aspect of this always did arise in connection with alignment. One generally does not observe an identical match of a significant number of residues in the whole sequence, but between fragments in the same order in both sequences. For example, sequences a x b x c x d x e and A X B X C X D X E are homologous because a shows significant homology with A, b with B, and so on, but with intervening sections x of variable length from zero to of the order of 10 residues. These sections x are the insertion/deletion regions. The remaining fragments a, A, b, B, etc. may also be of the order of ten residues. Bearing in mind the significant degree of residue alignment that can be expected by chance, alignment of the sequences to optimize the degree of homology, which may now describe as identification of homologous fragments, is often problematic. Alignment is helped by the idea of insertion/deletion regions being surface loops, while the remaining fragments a, A, etc. correspond to "core fragments". The surface loops tend to be predicted as loops in secondary structure predictions, tend to be more polar, and show much less homology with the loop in the corresponding location in a homologous sequence because of the enhanced probability that any substitution, insertion, or deletion mutation will be accepted at the surface. In contrast, the core fragments tend to be more hydrophobic, are predicted to be helices or β-sheet strands, and are more conserved in evolution leaving the degree of significant homology which leads to their identification. Secondary structure prediction and consideration of the average polarity in a region thus often help the alignment.

One may point out also that insertion/deletion regions are often involved in the acquisition/loss of exons during the course of evolution and consequently they cannot be considered only as a surface loop region. They even can be considered as self-consistent interchangeable structures or motifs which will be considered more extensively below.

A. Growth Factor Receptors, Protein Kinases, and Oncogene Products

Besides the previously mentioned identities between the $P28^{sis}$ product of v-*sis* and the growth factor PDGF there exist strong identities between the v-*erb* B product and the cytoplasmic part of the receptor of EGF.[4] EGF R is one of the best characterized growth factor receptors. It is a 175-kdalton glycoprotein which has an intrinsic protein-tyrosine kinase activity. Proteolysis studies show that about 115 kdaltons of the receptor on the outside of the cell form a heavily glycosylated EGF-binding domain. It constitutes the N-terminal side of a 1186 amino acid chain which commences with a 20-residue hydrophobic stretch presumably representing the signal peptide. In addition, residue 619 begins a stretch to residue 644 with all the properties of a signal peptide and which has reasonably been interpreted as the permanent transmembrane portion, joining a major domain on the outside surface and a major domain on the inside surface of the protein (see Figure 1). In very gross terms, the protein thus resembles a dumbbell in which a highly nonpolar connecting stretch, the bar of the dumbbell, spans the membrane. This seems to be a very common structure for a membrane-inserted protein. The initial signal peptide is followed by the sequence Glu-Glu-Lys-Lys, two negative and two positive sidechains which are, at least locally, a neutral configuration overall. The transmembrane portion, however, is followed by

Arg-Arg-Arg-His-Leu-Val-Arg-Lys-Arg-Thr-Leu-Arg-Arg

In charge notation, this sequence is + + + + 00 + + + 00 + +. This highly positive section presumably sits in stable contact with the inner negatively charged surface of the membrane. In contrast, the charged sequence immediately preceding the charged transmembrane section is essentially a mixture of positive and negative charges, interspaced with nonpolar residues.

Further examination of the sequence close to the highly positive post-transmembrane section reveals a compensating run of negative residues, then a compensating run of positives, and so on. However, preceding the C-terminal direction, there is a progressive mixing in that positive sections become interspersed by negative residues, and vice versa. That is, the plotted distribution of charges along the sequence resembles a progressively dampened oscillating curve, swinging from positive to negative but with the envelope converging to zero.

Computer molecular modeling of this region suggests a broad spiral or helix with 12 residues per turn (Figure 1). The surface charge distribution follows the distribution of positive and negative counterions near the surface of a negatively charged membrane, though it may be out of phase with it. This motif is thus not only compatible with the negative surface charges of the membrane, and with a distribution of surface charges which relaxes the electrostatic stress, but it is an electrostatic complement to the ion distribution near the membrane surface. However, it is likely that such a structure does not form a rigid structure like a compressed coil spring, but a more floppy one like a relaxed coil spring. Such a feature may have several roles. One is that it forms, with the nonpolar transmembrane segment, a structure which is complementary to a system comprising the membrane and its associated ion-rich solvent, and hence, unlike a signal peptide, prevents further threading of the protein chain through the membrane. The second, and related aspect, is that it forms a neck on which the intracellular domain is held. The third, is that this spring could undergo extension and compression with changes in the electrostatic environment including, perhaps, the ionic medium. Since it has been postulated that the protein is, or is associated with, a proton pump, this may be of control significance and might even represent part of the pump mechanism.

The internal (intracellular) domain has protein kinase activity and is, on the basis of homologies in sequence, a fairly typical globular protein with two main lobes. In other words, its architecture beyond the above coiled spring region is not outstandingly interesting as far as membranes are concerned, being a membrane protein motif only by virtue of its indirect linkage. It may, of course, have a surface pattern of charges on the side facing the membrane which is interesting, but these cannot be inferred from the sequence alone and detailed modeling would be required.

A 250-amino acid stretch of this internal domain has striking homology with the catalytic

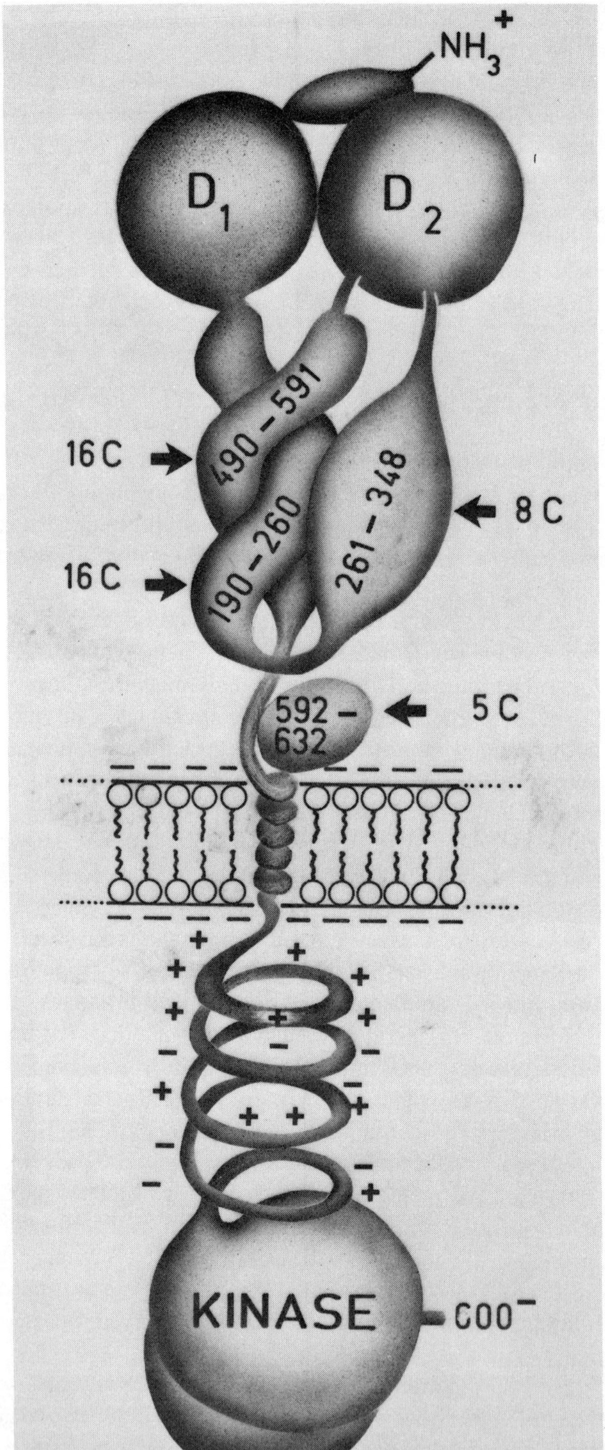

FIGURE 1. A molecular model of the EGF receptor at low resolution. The v-*erbB* oncogene will correspond to residue 551-1154 of EGF receptor with a small external fragment protruding the membrane and the cytoplasmic part with its two domains.

domain of the Rous sarcoma *src* gene family of protein-tyrosine kinases. Residues 551 to 1154 of the human EGFR show over 90% identity with the predicted sequence of the chicken v-*erb* B gene product.[3] The v-*erb* B gene product thus appears to be a truncated form of EGFR which lacks most of the external EGF-binding domain but which retains the membrane anchor domain. There is a somewhat similar identity (85%) of the cytoplasmic domain of the EGF receptor and the β subunit of the insulin receptor[9] and consequently to v-*erb* B product. Weaker identities (24%) or homologies (34 to 37%) on a stretch of 280 amino acids relate v-*erb* B oncogene product with *src,* Y73 avian sarcoma virus *yes,* Synder-Theilen feline sarcoma *fes,* Fujinami sarcoma *fps* gene products, and bovine cAMP-dependent protein kinase.[10] It is the C-terminal part of cAMP-dependent protein kinase which shows 21 to 24% of identical residues with *src* products.[11] There is also a good degree of homology between the cAMP-dependent protein kinase and Moloney murine sarcoma v-*mos* product and the cell division control protein CDC 28.[12] About 23% of identical residues are observed between v-*mos* (residues 1 to 374) and v-*src* products (residues 188 to 530).[13] This level of identity is similar to the one relating human myoglobin and human β-hemoglobin chain. Finally the region 1200 to 1400 of the 1434 amino acid residues of $P160^{gag\text{-}fms}$, the product of McDonough feline sarcoma virus, shows 35 to 45% of identical residues with regions of v-*fes* and v-*src* products.[14] A total of 436 amino acid residues out of the 526 residues of $P60^{src}$ are identical to $P90^{gag\text{-}yes}$ of 812 amino acid long.[15] $P120^{abl}$ of Abelson murine leukemia virus which comprises 917 amino acids shows 176, 166, 140, and 138 identical residues with *yes, src, fes,* and *fps* gene products, respectively.[16]

B. GTP-Binding Proteins and Oncogene Products

The *ras* family of oncogenes of Harvey and Kirsten murine sarcoma virus constitutes a third group after growth factor and protein kinase oncogenes whose protein product (P21) is a GTP-binding protein. Several reports have pointed to sequence homology between P21 proteins and various nucleotide-binding enzymes and regulatory proteins such as the β-subunit of bovine mitochondrial adenosinetriphosphatase,[17] bacterial elongation factors EF-Tu and EF-G, α-tubulin, adenylate kinase, and other nucelotide binding enzymes,[18,19] and signal-transducing G proteins especially the α-subunit of photoreceptor G protein.[20] The most promising homology is between P21 and the GDP domain of EF-Tu of which the three-dimensional structure at 2.7 Å resolution is known.[21] The primary sequence of P21 has 42% of amino acids identical or conservative equivalents in EF-Tu.[22] They are four principal regions of shared sequence similarity separated by stretches of nonaligned amino acids: Region I related to GTP binding (residues 15 to 29 in EF-Tu), region II related to Mg^{2+} ion binding (residues 79 to 87), and regions III and IV participating in guanine interactions (residues 99 to 108 and 128 to 140, respectively).

A careful comparison of nucleotide sequences showed that viral and cellular *ras* genes are nearly identical differing possibly by only one nucleotide base substitution at position 12 in the amino acid sequence when glycine has been found to be replaced by valine, serine, or arginine. For this reason, this sequence region is named transforming region and is likely part of the GTP binding site (see below). There are also good arguments that the c-*ras* protein of yeast is the G protein controlling the activity of adenylate cyclase.[23]

C. Steroid Hormone Receptors and Oncogene Products

The recent elucidation of the amino acid sequence of the human glucocorticoid receptor has revealed a remarkable homology of its carboxy-terminal half and the oncogene product $P75^{gag\text{-}erbA}$ from the avian erythroblastosis virus.[24] v-*erbA* oncogene strongly potentiates the effects of v-*erbB* by blocking the maturation of erythroblasts at an early stage of differentiation. Its amino terminal is synthesized as a fusion polypeptide with the viral gag product. Its N-terminal 1 to 398 amino acid revealed 85 identical residues with 22% amino acid identity overall with the last 387 amino acid residues of human glucocorticoid receptor. There is an even greater identity, 40%, between amino acid residues 421 and 481 of the receptor which corresponds to a cysteine/lysine/arginine-rich region. Nine out of ten cysteine residues of the receptor are conserved in this region of maximal homology with *erbA*. They could correspond to a putative binding site to DNA.

III. SECONDARY STRUCTURE PREDICTION AND MOLECULAR GRAPHIC MODELING OF ONCOGENE PRODUCTS

A. Principles of Secondary Structure Prediction

There are two ways of using secondary structure predictions: as a starting conformation prior to energy minimization (see Section IV) or as a suitable element to model the polypeptide chain to an already known protein structure to which it is homologous. This is usually done with the help of molecular graphics. We should first make some comments on secondary structure prediction methods. With the exception of some which use helix-coil transition data from experiments on artificial polypeptides, the vast majority of secondary structure prediction methods are statistical in nature. They depend on counting frequencies of events in tables of sequence-conformation relationships for protein of known structure. A crude example of a prediction approach would be one which starts with the observation that proline never occurs within a helical region in that database (except occasionally at the helix N-terminus). Thus, it would not be assigned as being within a helical region in the initial protein conformation from which minimization is to be performed. In practice, one develops algorithms and parameters which account for the observed sequence-secondary structure relation and which have proven value in predicting the conformations of proteins not in the data base.

Early prediction methods tended to classify each of the 20 amino acid residues as helix formers or helix breakers. The observation that no type of residue consistently adopted the same conformation was then overcome by simple rules (e.g., that at least four helix-forming residues had to occur in a run before helix would be formed). However, there is no reason to suppose that binary assignments of residues as helix formers or breakers reflect the real situation as demonstrated by helix-coil transition studies of artificial polypeptides. Therefore, Pain and Robson[25,26] assigned the 20 amino acids to a scale of helix-forming propensity making use of the information theory approach. This was to avoid attaching any prejudicial physical significance to the problem and instead to treat the sequence and conformation of proteins in the data base as two messages related by an unknown code. In order to reach objective conclusions, a Bayesian approach was combined with the information theory method. This formal basis of the procedure, and its applications, are discussed by Robson.[27]

The demonstrable power of the secondary structure predictive methods and their computational cheapness, have inspired a variety of approaches by many authors. However, emphasis is given here to the information approach. This is justified because the approach does not constitute a specific algorithm, but a general formalism for seeking optimal predictive algorithms. The predictive method of Chou and Fasman,[28] for example, can be shown to correspond closely to the information theory approach taken by Robson and Pain[26] when terms representing interactions between residues are neglected. The aspect of the information theory description as used in the predictive method of Garnier et al.[29] may be formulated as follows:

$$I(S_j;\ R_{j+m}) = \mathrm{Ln}(n(S_j,\ R_{j+m})/n(S)) - \mathrm{Ln}(n(\overline{S}_j;\ R_{j+m})/n(\overline{S}))$$

for large number of events. Then n (S_j, R_{j+m}) is the observed frequency (number of times) with which residue R at the position j + m in the sequence is found when the conformational state S (say α-helical) occurs at the position j in the sequence. n (S) is the frequency with which the conformational state S is found in the data base irrespective of the amino acid residue. $\overline{S}$ means any conformational state except S. Theoretically m should be any value of the number of amino acids in the sequence and the information values added. In practice the value of m has been chosen ± 8 so that

$$I(S_j;\ R) = \sum_{m=-8}^{m=+8} I(S_j;\ R_{j+m})$$

Here j is the residue whose conformation is to be predicted, S_j a conformation, and R_{j+m} the type of residue (alanine, etc.) m residues away down the sequence. A positive value of m implies a residue in the C-terminal direction of j, and a negative value a residue in the N-terminal direction.

The terms are negligible beyond m ± 8, though this cut off is somewhat arbitrary. The case m = 0 implies the type of residue j itself. This prediction is done for each residue in the sequence from j = 1 to j = n. This is then repeated for every possible conformational state S which one wishes to consider. For each residue the conformation S with which the highest information content is associated is the one predicted. In practice, it is useful to first subtract a "decision constant", dependent only on the conformation S, for the value of the information for each residue in each of its conformational states. This can be envisaged as extra information from circular dichroism studies, and it is not necessary to know the secondary structure contents exactly, but only to classify the proteins as helix rich, pleated sheet rich, etc. There is physical justification for this. For example, a protein rich in potential sheet regions will further stabilize pleated sheet formation by a cooperative hydrogen bonding between the strands. More importantly, long helixes are more stable than short ones.

The study by Garnier et al.[29] is still fairly typical in that four conformational states (right-hand α-helix, extended chain as potential pleated sheet, reverse turns, and aperiodic structure) are considered. In considering assignment of four types of conformational states, roughly 60% of residues will be correctly predicted as being in the experimental state. This is, however, over many proteins, and a few proteins show marked deviations from this. The procedure is readily coded in about one page of a high level computer language, and for short sequences can even be carried out manually. The computer program is available on request to the authors with a recent version in basic language for microcomputer.

B. Use of Secondary Structure Prediction in Predicting Tertiary Structure: Supersecondary Motifs and Domains

Now one can address the question of how these secondary structure features, or at least the common periodic secondary structure features α-helix and β-pleated sheet, come together in space. Here we are concerned with viewing helices and sheets as solid objects, and seek to find, for example, their preferred points of contact and the preferred relative orientation.

In theory, the problem of understanding the association of these structures is the same as that of understanding a dimerization of two molecules including the case of formation of an enzyme substrate complex. This immediately implies the idea of the progressive folding up of a protein molecule by formation of secondary structure and then association of these into *supersecondary structure* features. From a nondynamic viewpoint, however, we may simply regard supersecondary structure features as a hierarchical level of structure representation between the secondary and tertiary levels.

There is of course no *a priori* reason why the association should be simple to describe just because it is precise. A substrate-enzyme interaction, or the fit of a key to a lock, does not depend on regular surface features of the interacting components. Regular surface features on helices and pleated sheets do however become manifest if we build models of them with, say, all-alanine or all-cysteine sidechains. Then, a pattern of grooves is apparent which implies a limited number of possible mutual orientations corresponding to optimal meshing of the grooves. These suggest preferred orientations for which there is some evidence when we examine the distribution of orientations in known structures. In reality, we might expect the vast variation in sidechains on helices and sheets, and sidechain flexibility, to "roughen" the groove pattern, causing the simple packing rules to be lost. Some analyses take some account of the sequence or at least the residues at point of contact. As yet, the status of the art is such that it is only possible to make tolerable predictions of packing arrangements for helix-rich proteins; however, the developments in considering pleated sheets are extremely promising.

In terms of connections between secondary structure features, much pioneer work has been done in this area by Richardson[30] allowing one to compare connections between secondary structures in different proteins, show underlying similarities, and thereby to suggest possible evolutionary relationships as well as allowing a simple description of gross folding patterns in proteins.

Studies on preferred orientations as described above have shown the tendency for adjacent structure features in the sequence to lie roughly antiparallel. Preserving at least a crude representation of relative orientation is also important in describing connections, so this aspect should be implicit in a tentative notation. Let us use the terms a, b, x to represent a strand of α-helix, a strand of pleated sheet, and a strand which will typically be coil but which may contain a helix or sheet strand if we do not wish to specify this and so emphasize the basic similarities. Writing two adjacent symbols will thus imply that they are antiparallel unless specified otherwise. Common "motifs", or components of motifs, are aa which represent consecutive antiparallel helices and bb consecutive antiparallel pleated sheet strands; in both cases, the backbone describes a rough U-shape and the chain leaves in the opposite direction to the one in which it enters the motif. There must be at least a very short connecting aperiodic region, but this is implied rather than represented in the notation. To represent two consecutive parallel helices, we write axa. Since the first and second strands are antiparallel, and the second and third strands are antiparallel, the parallel orientation of the helices is implied. In this sense, x sometimes has the significance of a "dummy" strand used to represent parallel relationships, but this is reasonable since the strand between helices and sheets in axa and bxb, respectively, must generally be longer, and more apparent, than in aa and bb. Since we are only concerned with "roughly parallel" or "roughly antiparallel" in this kind of description, precise information about orientation will be lost for items far apart in the backbone in complex structures. This is less so in pleated sheet motifs where the hydrogen bonding between strands better preserves a more parallel or antiparallel arrangement, though it should be noted that, pleated sheets in globular proteins are not generally planar; neighboring strands show a slight angle between their axes causing the sheet as a whole to appear twisted.

In discussing pleated sheets, we shall assume that successive b strands are hydrogen bonded even if there is intervening non-beta structure, as in the common motif bab. If we wish to state that this is not the case, we may write b/ab or ba/b, depending on which of the two possible a-b associations is weaker. The motif bab is very common and has handedness. That is, because the β-pleated sheet is significantly twisted, there is a shorter path for the helix strand over one side of the pleated sheet than the other, and this path is most generally preferred in proteins. Idealized sheet motifs are bbbb . . . which is antiparallel pleated sheet, bxbxbx . . . which is parallel pleated sheet, and bababa . . . which is a ROSSMAN-fold pleated sheet characteristic of proteins binding nucleotides (see examples in Table 1).

A numbering scheme can be proposed as follows. A signed number is inserted in brackets

Table 1
CHARACTERISTIC CHAIN FOLD IN PROTEINS

aa

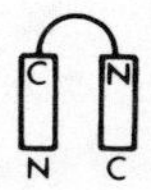

Up-down helixes roughly antiparallel as in utero-globulin, hemerythrin, and cytochrome c′

axa

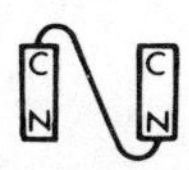

Helixes roughly parallel with β-strand connection

$a^c = a(+2, +3)\ a(+1, +2)\ a(-1, -2)\ a(-2, -3)$

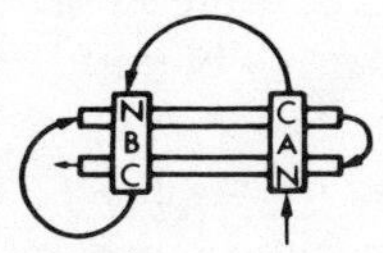

Hydrophobic contacts between touching helixes as in globins, cytochrome c, thermolysin, papain, and T4 phage lysozyme

$bb = b^2$

β-strand folds back on itself to form antiparallel pleated sheet very common, e.g., pancreatic trypsin inhibitor

$bbb = b^3$ (egg lysozyme)

Two inversions forming an S are common

bxb (crossover β-sheet)

Parallel pleated sheet with intervening strand: helix (bab) or coil (bcb) or strand involved in other pleated sheet (bbb). Often the connecting x (a, b, or c) strand passes over the β-strand to β-strand hydrogen bonds but in subtilisin and glucose phosphage isomerase the x-strand passes underneath

babab (Rossman fold)

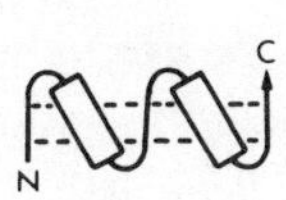

Two fused bab units in many proteins, especially those binding nucleotides

$b^G = b(+3)b^2/b(-3)$

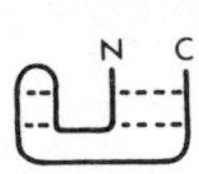

Very common as segment of β-sheet barrel forming a Greek key β-sheet as trypsin, pyruvate kinase, prealbumin, plastocyanin, immunoglobulin C, staphylococcal nuclease, Cu, Zn superoxide dismutase

$C = b(+5)/b^5/b(-5)$

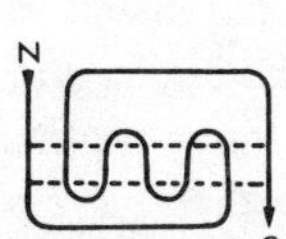

Contains embedded bb hairpin β-sheets or jumping cracker Greek key β-sheet as for example in immunoglobulin V

$j = b(+5)/b(+3)/b^3/b(-3)/b(-5)$

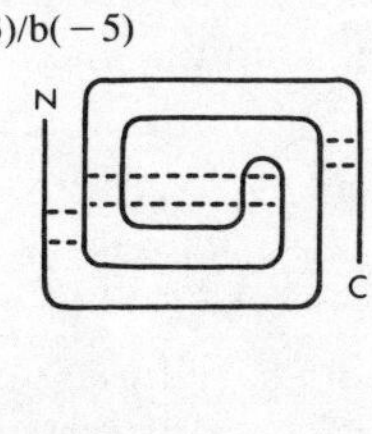

Contains "nested" Greek key bundles (jelly roll) as in tomato bushy stunt virus

$bb(+4)b^2a/b(-4)\ (ab)^2$

Complex cross-over β-sheets. The formula of hexokinase sheet is shown here but very common, e.g., flavodoxin, glyceraldehydephosphate-dehydrogenase, subtilisin, dihydrofolate-reductase, adenylate kinase following a strand to show further associations. For example, a pleated sheet may be bent into a cylinder hydrogen bonding the first b strand to the last: the so-called "barrel" which, however, resembles a cylinder with a "nipped waist" due to the twist of the sheet. Ideal barrels have the general formula $b^n b(-n)$ or $b(+n)b^{n-1}b(-n)$, where, e.g., h b^n simply implies b written consecutively n times, and $b(-n)$ means that the strand preceding the bracket is hydrogen bonded to the strand $-n$ strands away, that is, n strands to the left as the formula is expressed. In this example, $n + 1$ is the number of strands in the barrel. In fact, such ideal barrels are probably rare. The serine protease barrels are more complex. The structure in triosephosphate isomerase closely approximates a barrel of the type $ab(+2n)$ $(ab)^{n-1}ab(-2n)$ or more specifically $ab(+14)$ $(ab)^6ab(-14)$, where we have a barrel of antiparallel beta strands as the repeated motif bab. Note that the counting describing connections counts all strands, including a (and x if present).

Related to the general barrel class $b(+n)b^{n-1}b(-n)$ is the general Greek key class of pleated sheets $b(+n)b^{n-1}/b(-n)$ where the last strand is hydrogen bonded to the first but not to the strand immediately preceding it in the sequence. These are thus barrels unfolded into the plane by the breakage of one connection, and resemble motifs found on Greek pottery. The basic Greek key is $b(+3)b^2/b(-3)$ for which a barrel would be sterically impossible (hydrogen bonding NH and CO groups would have to be bent to right angles to each other). In that sense, a typical Greek key is an attempt at a barrel which fails because of two few strands. However, they may appear as barrel components. A helix strand substituted for a beta strand is also possible, as in $b(+4)b^2a/b(-4)$. This is encountered in many proteins in association with a ROSSMAN type fold bababa . . . , for example, $b(b + 4)b^2a/b(-4)(ab)^2$ in hexokinase, flavodoxin, glyceraldehyde phosphate dehydrogenase, subtilisin, dihydrofolate reductase, and adenylate kinase (Table 1).

The general Greek key class is related to the barrels on the one hand, and on the other to the general jelly roll class $b(+n)/b^n/b(-n)$. If a general Greek key is $b(+n)b^{n-1}/b(-n)$ or equivalently $b^n/b(-n)$, then the general jelly roll class must embed a Greek key. Hence $b(+5)/b^5/b(-5)$ in Immunoglobulin V resembles two fused Greek keys (Table 1), it is technically a "Greek keys beta sheet". Let us take the b^5 motif in this formulation and replace it by the basic jelly roll $b(+3)/b^3/b(-3)$. This has the same number of strands and, as this is an odd number, the entering and exit strands are in the same parallel orientation allowing similar connections. We then obtain $b(+5)/b(+3)/b^3/b(-3)/b(-5)$; the "true" or "classic" jelly roll found in tomato bushy stunt virus protein.

Proteins in which we have two or more regions engaged in their own pleated sheet or other supersecondary motif are common (the separate motifs often correspond to separate domains, see below). They are readily represented simply by increasing the length of the formula, which generally simply means writing the formula more than once since the numeric description of the connections is relative rather than referring to the absolute strand number. Less trivial is the case when there are two pleated sheets with a complex weaving of the backbone from one sheet to the other, so that the chain does not remain in any one strand for any great distance. Examples include the immonoglobulins, prealbumin, superoxide dismutase, and concanavalin A. The serine protease barrels also present this difficulty in that the chain jumps from one side of the barrel to the other and also exemplifies two independent components in that the two barrels are quite distinct.

Descriptions of supersecondary structure motifs which are solely or mainly α-helix in content are only marginally more difficult in that the associations of strands primarily involve hydrophobic contacts, which may allow more associations with a given strand. There is also the relaxation of the need for strongly associated strands to be close to parallel or antiparallel; indeed, successive helices may lie at approximate right angles and associated helices may lie approximately at a right angle. At least in a small structure, this information may be represented by a′, which signifies that successive strands lie at a right angle. We have stated that aa is common; the next most complex form is the Greek key helix bundle

$$a(+2,+3)\ a(+1,+2)\ 'a(-1,-2)\ a(-2,-3)$$

which resembles two U-bend aa structures one on top of the other (Table 1). This structure has handedness which could be represented by two kinds of right angle symbol, but we shall not pursue this here. The extensive cross connections also mean that there is a stronger affinity between this kind of helix Greek key and a helix barrel than in the analogous case of the pleated sheet motifs. For the helix barrel there is some resemblence to four helices with their axes at the parallel edges of a box but twisted sharply round the parallel axis through the center of the box. That is, they somewhat resemble a four-stranded rope.

In supersecondary structure we are looking at features between the secondary and tertiary level from the viewpoint of the secondary structure level. We may also look down from the tertiary structure level, and in practice this means looking for parts of the globular structure of a protein which themselves look like distinct component structures, or *domains*. Such structures are certainly found, typically as compact, rather hydrophobic clusters of residues formed by a local bunching up of the polypeptide chain. They may be identified and characterized on this basis though the problem is frequently in knowing where to stop, to avoid resolving domains into further domains. In more obvious cases, domains resemble protein subunits, except that they are not separate molecules but rather connected by the continuity of the polypeptide backbone. Since with supersecondary structure we look up from the secondary level, and with domains we look down from the tertiary level, it may well be asked what the difference is between domains and the kind of supersecondary structure features described above. In many cases, the answer would be not much. A serine protease essentially represents two domains with the active site represented by the cleft between them, a fairly typical situation for an active site. Each such domain is a pleated sheet barrel, with sidechains "added" to the backbone, and extensions of the polypeptide chain at the ends. Typically, a domain may be primarily a single supersecondary structure motif, or a set of smaller supersecondary motifs, sometimes combined in an elaborate way which may still be described approximately by the kind of notation used above.

Domains typically consist of 50 to 150 residues. Thus stating that a helical strand, a β-sheet strand, and an intervening coil strand are each of the order of ten residues should provide a rough picture of the size of the supersecondary structure motif or set of such motifs required to makeup a domain. Alternatively, supersecondary structure motifs would constitute substructures or *modules* within domains. Go[31,32] has developed a computer algorithm on the basis that they represent compact structural units or least extended conformation in proteins of known structure. The interest in this approach is that very often these modules of compact structure match the span of amino acid sequence corresponding to an exon. Split gene structure in eukaryotes has been recognized as an important structure for tracing molecular evolution of proteins.[33,34] It is too early to pronounce that these modules coded by exons do possess a structural autonomy but they do look like pieces of a puzzle, 20 ± 10 amino acid residue long which in Go's view may have been used in the course of evolution of eukaryotes to build up novel proteins.

C. Application to Molecular Modeling and to Oncogene Products

Following the pioneering work of Lewinthal, molecular graphics has made considerable technical progress. It now allows one to build up on a television screen complex molecules such as proteins, to rotate them in real time, or to modify them at will and still being able to extract the coordinates of all the atoms. On the principles of homologies in amino acid sequence or in supersecondary structure motifs several attempts have been made to model the structure of oncogenes either partially or totally. Sternberg and Taylor[35] have proposed a molecular model of the ATP binding site of v-*src* product. They identified a sequence of amino acids of the type $G\ X\ G(X)_2G(X)_nVA\ X\ K$ common to various oncogene products:

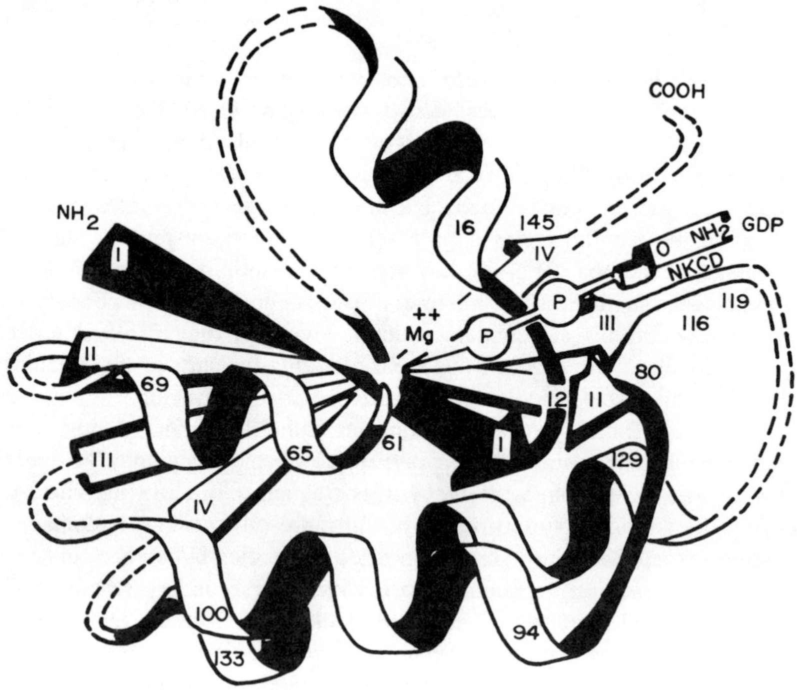

FIGURE 2. Model for the tertiary structure of P21. Broken lines represent regions in which sequence homology with EF-Tu is insignificant. (From McCormick, F. M., Clark, B. F., LaCour, T. F. M., Kheldgaard, N., Norskov Lauretsen, L., and Nyborg, J., *Science*, 230, 78, 1985. With permission.)

v-*src*, v-*erbB*, v-*fps*, v-*yes*, v-*abl*, v-*fms*, v-*mos*, and to the nucleotide binding regions of five dehydrogenases of known structure. Making use of X-ray coordinates of glyceraldehyde-3-phosphate dehydrogenase they proposed a model of folding for the fragments 268 to 297 and 399 to 408 of the oncogene product of v-*src*. These fragments would by predicted in terms of secondary structure as abaxb, i.e., of a Rossmann fold type.

A molecular graphic model of P21, the product of the *ras* oncogene has been recently developed by McCormick et al.[36] The amino acid sequence of P21 has been aligned with the bacterial elongation factor EF-Tu so that 42% of P21 amino acids have identical or conservative equivalents in EF-Tu. They predicted secondary structures according to Chou and Fasman[28] and Garnier et al.[29] Then from the X-ray coordinates of EF-Tu of the conserved amino acids and the secondary structure prediction of P21 they generate a model of P21 on the graphic system (see Figure 2). This model predicts the fragments of P21 sequence responsible for guanine nucleotide specificity and for binding and hydrolysis of guanosine triphosphate (GTP). This included all known biochemical properties of binding of GTP of EF-Tu and P21. From this molecular model the replacement of glycine at position 12 by a valine would clearly alter the binding of GTP and this is considered by the authors at the origin of the tenfold drop of GTPase activity of P21. This reduced GTPase activity would maintain P21 in an active (GTP) state much longer than the proto-oncogene. This would be at the origin of the transforming property of the oncogene since it is observed that the replacement of Gly at position 12 with any other amino acid, except proline, activates the transforming potential of P21.[37]

IV. CALCULATION

The calculation of peptide and protein conformation rests on the basic assumption that this conformation is of least free energy. Then arise two major problems: how to determine the free energy of the system and how to locate the conformation of least energy. There is not, for the moment, an exact solution to these two problems, only approximations can be made and tested. Rather than do an exhaustive presentation of the approximations used by different authors (see Robson and Garnier[38]) we will discuss those used for calculation on oncogene products.[39,40]

To locate the lowest minima one is faced with the difficulty that if a protein has n rotable bonds, then there is an n-dimensional potential surface for the protein (even neglecting changes in bond lengths, valence angles, and the solvent conformation). Worse still, this potential surface contains what may be millions of minima from which the deepest (''global'') one is to be located. This is called the ''multiple minima problem''. To tentatively solve this problem, time saving computer tricks or heuristics have been developed. The simplest of those is the choice of the *starting conformation*.

For the decapeptide of the transforming region of P21 of v-*Ha ras* oncogene the potential surface is much more simple than would be the whole oncogene product, however, here the minimization was started from a cluster of hydrophobic residues. The decapeptide was built-up progressively by adding di- or tripeptide units. These units were previously energy minimized and the peptide was energy minimized again on all dihedral angles after this addition. All interatomic interactions were calculated using the potential function of the empirical conformational energy for polypeptides and proteins (ECEPP) program.[41] The minimization procedure was a gradient method which searched for the nearest local minima (see discussion below). The calculations showed that the most favorable form of c-*Ha ras*-P21 decapeptide exists when glycine 12 is in a left-handed bent conformation. No other amino acid was found to adopt this conformation and thus the bladder oncogene peptide containing valine at position 12 would have a markedly different conformation.

For the far more complex calculation performed on the v-*sis* viral oncogen product, $P28^{sis}$, which comprises 160 amino acid residues, Robson et al.[40] used a secondary structure prediction of backbone dihedral angles ϕ and ψ. A combination of three methods was used: (1) the procedure of Garnier et al.;[29] (2) a method based on that of Lim; and (3) a new procedure similar to that of Garnier et al.,[29] but extended to 81 rather than 4 states, each state consisting of a 40°C by 40° area of the ϕ and ψ map. Methods (1) and (2) were the most satisfactory of those objectively tested in predictions of three states (helix, pleated-sheet, and coil).[42] The overall procedure predicts backbone bond rotation angles ϕ and ψ which are a ''best compromise'' of the methods, as determined from proteins of known structure. This also allowed a degree of confidence (a value between one and four) to be assigned to each residue on the basis of the agreement between the methods (e.g., one if all methods agree, four if all methods disagree). The ϕ, ψ angles finally assigned were those which were most common for that type of residue in the predicted 40° × 40° angle range.

The second stage of the calculation involves *the minimization of the energy*. At the level of complexity even for peptide the only tractable energy functions that can be used are simple analytical (energy) functions, (i.e., potential functions) by which the energy may be estimated from atomic (primarily nuclear) coordinates. This bypasses the need to consider the electronic behavior of the system in the quantum mechanical sense. The term ''empirical'' arises from the extensive use of adjustable parameters which are ideally selected to give agreement with a sufficiently comprehensive body of experimental data. However, ''empirical'' is sometimes a misnomer because of the increasing use of *ab initio* calculations in providing those parameters instead of experimental data. This is currently the most important application of the *ab initio* approach in peptide work.

Empirical potential functions need only reproduce the relative energies of different conformations since this is all that is required for a calculation of conformational behavior. This is fortunate, since simple functional forms could not reproduce the behavior of electrons which determines the absolute energy of a specific configuration of nuclei. Empirical potential functions for interatomic interactions are generally of the type

$$E_{ij} = E(r_{ij})$$

which is to say that they describe the interactions between every pair of atoms i and j at a separation r_{ij} between the nuclear centers. They are also made to depend on the nature of the two atoms, e.g., sp^3 carbon with sp^2 carbon. The assumption of pairwise additivity implies an approximation to the total energy E(q) being given by the sum over all unique interactions for a set of Cartesian coordinates q

$$E(q) = 1/2 \sum_{ij} E(r_{ij}),\ E_{ij} = E_{ji}$$

with neglect of cross terms which would represent any special effects of three-body interactions, etc.

The analytic form of empirical potential functions E(r) is based on generalized quantum mechanical calculations. That these forms do not in general resemble $E = e^2/r$ is due to the fact that they represent the gross averaging over electronic behavior required to treat atoms as if they were single particles. In practice, each atom is typically treated as a spherical electron gas in order that the appropriate averaging can be easily carried out. For convenience, the functional representation is split into additive terms, each representing energy associated with some aspect of interatomic behavior. These are as follows:

1. A repulsive energy at short distances, due to electrostatic repulsion of the electrons. This is generally taken as proportional to r^{-n}; where the choice of n is problematic it was 9 in calculation made by Robson et al.[40] and 12 in Pincus et al.[39] Alternatively, an exponential form is used, which has a better theoretical justification except for close approaches. Strictly, this contribution is nonadditive, but the large power on n means that the energy rises very rapidly over a small change in r. Thus, it is this term which invokes comparison with the hard sphere model which implies the discontinuous function and the notion of a specific van der Waals's radius for each atom, the sum of these radii for an interaction pair being R_{ij}.
2. A dispersion energy due to the mutual perturbation of the electron clouds so that there is a compensating movement of charge. This compensation (correlation) implies that the interaction is always favorable, contributing a negative energy. At any moment of time, one may envisage as a transient positive ''end'' of one atom lying adjacent to a transient negative ''end'' on the other. Actually, the system is in oscillation with the positive and negative ends exchanging at a frequency of the order of 10^{-15} sec, characteristic of electronic motions in atoms. The term is invariably taken as being proportional to r^{-6}, characteristic of an induced dipole-dipole interaction. However, over longer distances the electrostatic field cannot keep up with the electronic oscillations, and the effect is that the resulting delay in response of one cloud to the other causes the functional form to go proportional to r^{-7}. All this is not too important in routine conformational calculations, since the favorable energy contributed by this term is rather weak (about 0.1 kcal/mol^{-1} as atoms begin to touch), and dies away with distance. It is important to note that the real reason for retaining any dispersion term is that it adds greater flexibility to the description of the repulsive energy. In particular, since this term becomes more negative with smaller r, while the repulsive term becomes even more rapidly positive, the sum of the two is a function with a minimum. When this minimum is small, the general appearance is even more like the hard sphere model. Probably, the dispersion contribution is almost, though not exactly, additive.

3. A coulombic energy proportional to r^{-1} due to the fact that atoms in molecules are rarely netly neutral and appear, at least at a distance, as single points of charge (monopoles). The partial charges carried by the atoms are, of course, due to their different electronegativities (electron pulling power). Typical values are 0.1 to 0.5 electron units (e.u.). The energy may be positive or negative, depending on whether the charges are of identical or opposite sign. Magnitudes of up to 30 kcal mol^{-1} are not uncommon at the distances of "touching" atoms, and it is thus fortunate that this contribution can be considered as additive.
4. An induction energy representing an induced dipole in one atom from a permanent monopole on the other is clearly an intermediate case between the dispersion and coulombic effects. Appropriately, it is predominantly proportional to r^{-4} and has an intermediate energy. However, it is nonadditive and a detailed analysis reveals complex behavior. Perhaps for this reason alone, this term is usually neglected. It is hoped that the other terms have sufficient flexibility to reproduce together the required energy surface and this means that some unrealistic behavior may have to be implemented in the other terms to allow for neglect of induction.
5. Resonance energy arises from a purely quantum mechanical phenomenom relating to a change of degeneracy in electronic behavior on introducing asymmetric interactions via an approaching atom. Like the induction term it is relatively complex, certainly nonadditive, and almost always neglected.
6. Valence energy arises from distortions of geometry of covalently bonded atoms from equilibrium values. This applies to interactions between covalently bonded atoms A–B, and may include the A . . . C interaction in A–B–C as a function of the A–B–C valence angle. Valence energy may be neglected in the rigid geometry approximation, but should be included in any accurate calculation. In the *harmonic approximation* the valence energy is held to rise proportionally to $(r-r_o)^2$ (bond lengths) or $(\theta - \theta_o)^2$ (valence angles) where r_o and θ_o are "equilibrium" values. Then valence energy can also be expressed as a distance function between A . . . C in A–B–C. When $r = r_o$ or $\theta = \theta_o$, which is usually the case when the rigid geometry approximation is used, then the valence energy is set to be zero since only relative energies are of interest. Above, we state "equilibrium" in quotes because r_o or θ_o will often be estimated, as some standard value from *ab initio* calculations or X-ray studies. The favored values are of course the equilibrium ones in any specific structure.
7. Intrinsic rotational energy and anisotropy. The intrinsic rotational energy is in our view something of a myth. It represents the remaining energy which must be added as a function of a bond rotation angle when the above contributions do not give a satisfactory account of the way the energy changes with rotation. It is typically added as an intrinsic rotational potential of the form:

$$E_\theta = K_\theta(1 - \cos n(\theta - \theta_o))$$

where θ_o is an angle at the minimum of the potential, n the number of minima (typically 2 or 3), and the constant K_θ is the maximum value of the energy (the minimum value is zero). The value θ_o will usually correspond to a staggered conformation with the maximum value of E_θ corresponding to eclipsed conformations. With $n = 3$, there are three equal maxima (of height $E_\theta = K_\theta$) and three equal minima ($E_\theta = 0$).

In our experience, such a potential is namely a correction factor which can often be avoided if attention is given to orbitals, not simply atoms. That is, orbitals are represented as "dummy atoms". Some of these orbitals (core orbitals) will correspond to existing atoms in normal calculations, while others, such as lone pair orbitals, are represented as further dummy atoms bonded to the core orbitals. These can be found by quantum mechanical methods, and the centers of the dummy atoms are placed at the dipole centroid of these orbitals. These interact with each other and with core orbitals with the same kind of potential functions as for atoms. After all, the main difference is that our orbitals are ''atoms'' without a nucleus, which affects only the charge we assign to the orbitals. For most purposes it suffices only to assign a coulombic term (see (3) above), though when the orbitals approach closely, van der Waals repulsive terms and the dispersion term should be included. For most purposes it also suffices to neglect the orbitals concerned with covalent bonding.

Important exceptions are pi orbitals, and C–H sigma bond orbitals in alphatic CH, CH_2, CH_3 groups. For the latter, van der Waals repulsive terms and the dispersion term are required, since in the bond system CH_2–CH_2, for example, a sigma orbital in the first CH_2 approaches close to a sigma bond in the CH_2 in all conformations.

This approach is the orbital force field (OFF) method of Robson and Platt which has further advantages since it introduces greater electronic detail into empirical modeling of molecules. It also allows for atomic *anisotropy,* i.e., the atoms are not always spherically symmetric, but have bridges represented by the orbitals. The alternative approach, of using intrinsic rotation potentials is nonetheless faster and simpler when there are complex quantum mechanical interactions responsible for preferred bond rotation angles. One such effect may be the anomeric effect as in groups such as –O–C–O–, but this is not relevant to proteins and peptides.

Note that any rotational potential required for N-C_α and C_α-C′ rotations of the peptide backbone will be rather weak, and can be neglected.

Thus, one usually writes the total energy as

$$E_{ij}\text{total} = E_{ij}\text{repulsive} + E_{ij}\text{dispersion} + E_{ij}\text{coulombic} + E_{ij}\text{valence}$$

The usual potential function used in the case of rigid geometry approximation is

$$E_{ij}\text{total} = A_{ij}r^{-9} - B_{ij}r^{-6} + C_{ij}r^{-1}$$

where A_{ij}, R_{ij}, and C_{ij} are parameters dependent on atom types i and j.

The precise form and parameterization of these terms are important problems and an example of parameters is given in Table 2.

In calculations performed by Robson et al. on v-*sis* oncogene product a further approximation was made by representing each amino acid residue by 4 ''dummy'' atoms.[45] This reduces significantly the computing time which depends on the number of pair interactions to be calculated; they are (n/2) (n − 1) for n atoms, i.e., roughly proportional to n^2. The conformational space was also simplified by coupling the two ϕ and ψ angles of each amino acid residues to one variable γ taken as the angle of a vector with the diagonal axis of the ϕ and ψ map, the extremity of the vector describing an ellipse.[45] In order to increase the efficiency of the calculation of the empirical conformational energies, a ''zero-minimum'' interaction energy was used. This is so called because the value of the energy function is zero at its minimum, i.e., the energy function is always positive. This allows to reject a conformation when its energy is higher than a previous one (see Simplex method below) prior to the calculation of all the interactions since supplementary terms will only contribute to increase the energy function. Applied properly, it can be proven that relative energies of conformers are *not* affected by the algorithm.

The minimization of the energy function is the other important step in this second stage. From one point of view, minimization is a sampling procedure as are molecular dynamics and the Monte Carlo approach. However, it does not by itself make any reference to statistical mechanics precisely because it concentrates on the calculation not of average properties,

Table 2
HAGLER-HULER-LIFSON PARAMETERS FOR USE WITH THE 9-6-1 POTENTIAL FUNCTIONS

	A_{ii} (kcal/mol A^9)	B_{ii} (kcal/mol A^6)	C_i Electron units (charge on electron = − 1 e.u.)	
H	445	15	+0.11	Aliphatic
C	38,900	1,230	(chose to neutralize CH_n)	
C′	12,500	335	+0.46	Carbonyl
O′	45,800	1,410	−0.46	
N	36,900	2,200	−0.26	Amide
H	0	0	+0.26	

These parameters are used in the equation

$$e_{ij} = \frac{A_{ij}}{r^9} - \frac{B_{ij}}{r^6} + \frac{C_iC_j}{r}$$

where $A_{ij} = (A_{ii})^{0.5} (A_{jj})^{0.5}$ and similarly for B_{ij}. They apply to the backbone peptide group derived by Hagler et al.[43] and Hagler and Lifson[44] from crystals of peptide-like molecules. They have been refined and recalibrated for both backbone and sidechain atoms by Robson and Platt.[52]

but of most probable conformation. This is done by minimizing the energy as a function of conformation and discovering the most probable conformation as that of least energy.

Strictly, of course, it should be the free energy which is minimized as a function of conformation. What is done in practice is to make the reasonable assumption that a minimum in the potential surface must correspond to a minimum in the free energy surface. Then, one can make the further, and much less reasonable, assumption that the minima are of similar shape and width, so that the minimum of least energy also corresponds to that of least free energy. When there is no justification for this (and there rarely is), the latter assumption should not be made and the free energies at the local minima should be calculated and compared. The several methods of calculating free energy all necessarily make the prediction that a broad minimum of high potential energy can still have a lower free energy than a narrow minimum of low potential energy, providing the potential energy difference between them is not too great. The methods include the obvious one of integrating in the vicinity of each minimum to evaluate the local partition function Z, calculation of free energy as a function of second derivatives, and calculation of free energy based on vibrational frequencies.[46] Thus, statistical mechanical reasoning can and should be brought in at a second stage, to analyze the minima located by minimization of the potential surface alone. However, this further step is a fairly recent and rare innovation in peptide conformational analysis. It should also be said that it is theoretically possible to evaluate the free energy from the potential surface at all points, not just in the vicinity of the minima, and that a judiciously programmed procedure could provide a cheap method of minimizing free energy from the outset.

Minimization of the potential energy as a function of conformation is widely used. It is the basis of the computer simulations of protein folding and of the energy refinement of protein structures deduced by X-ray analysis. A variety of program packages are available for doing the minimization part of the calculation; most often these are general packages for which the user must write his own function for calculating energy from conformation. The variety of packages reflect the variety of mathematical approaches to determine the minimum value of a function. However, these basically fall into two types.

The first type is represented by gradient methods. Although the general principle is to locate that point for which the first derivatives with respect to conformation are zero and the second derivatives positive, some procedures do not require derivatives to be provided. The steepest descent method depends on one of the most approximate expansions, but has still proved popular in conformational work because of its convergence properties. Convergence is, however, slow. The Newton-Raphson procedure is a more sophisticated expansion requiring the evaluation of the second derivatives; there is a recent renewed interest in second derivative minimizers. The Fletcher-Reeves methods avoid explicit evaluation of the second derivatives, gaining information about curvature of the energy surface by assuming a quadratic form. Procedures due to Davidon[47] and Fletcher and Powell[48] take advantage of features in both Newton-Raphson and Fletcher-Reeves methods. These and subsequent techniques[49] have a considerable degree of efficiency and sophistication, but all suffer from the fact that the procedures stop at the first minimum located. Indeed, to exit from a minimum would be contrary to their basic philosophy.

The second type, represented by search methods, can also be fairly sophisticated as exemplified by the SIMPLEX procedure (Nelder and Mead[50]). It is well suited to minimizing energy as a function of all bond rotations rather than, x, y, z coordinates. A "simplex" (n + 1 points in n-dimensional space) samples the function value at its points and on this basis decides where to move its highest value point in an attempt to find a point of even lower energy. This involves operations on the SIMPLEX such as reflexion, then expansion or contraction of the probe point through the centroid of the remaining points. It would not be too misleading to speak of the alternative gradient methods as resembling a ball rolling downhill, while the SIMPLEX method resembles a blind octopus groping with its tentacles for its hole in a coral reef. While this understates the sophistication of the SIMPLEX method, it conveys something of its robustness. It is not confounded by discontinuities in the derivatives or in the energy function itself, and is indeed ideally suited to handling artificially imposed constraints, either on the conformational variables or on the energy or some other function of the conformational variables. Further, it has a chance of escaping from shallow minima if a probe point encounters a region of even lower energy. The chances of this happening depend on the depth and shape of the minimum encountered and the width of the energy well, and also on control parameters for the SIMPLEX procedure. By allowing a new point to be generated at a considerable distance from the centroid of the simplex, the chances of encountering new minima are obviously increased. In practice, modifications of the original SIMPLEX procedure may be required to treat particular problems more efficiently (e.g., COMPLEX). The price of the considerable robustness of the SIMPLEX approach is its slowness.

On the one hand, therefore, gradient methods locate local minima rapidly, but are less robust and run into trouble if the function has certain properties (such as the possession of discontinuities). On the other hand, the SIMPLEX approach can attempt to handle most functions and may permit escape from local minima thus increasing the chances of locating the global minimum. There is much to be said for combining the two for peptide-protein work. Thus, Robson et al.[40,45] employed a double minimization in which the gradient method would switch to a SIMPLEX mode if it became trapped in a minima or ran into trouble, and conversely the SIMPLEX would switch back to a gradient method when the vicinity of a new minimum was located. The switching was entirely automatic, being governed by diagnostic variables generated by each minimizer and by the current degree of convergence. This avoided the need to implement special procedures for escape from local minima (see, for example the "thermalization" method of Levitt and Warshel[51]).

Stage three of v-*sis* calculation was a compaction step to mimic solvent effect. A sigmoidal compaction potential was introduced to reduce the surface area of the whole molecule without introducing excessive internal stress and the concomitant formation of large energy barriers against internal movements:

$$C = S \sum_{ij} \frac{B_{ij} R_{ij}^2}{D^2 + R_{ij}^2}$$

Here R_{ij} is the distance between groups i and j; B_{ij} is a group-dependent parameter as in Reference 45; S is a constant (10^{-5}), and D is calibrated specifically for the molecule such that a typical protein density is obtained on compaction stage, but is subsequently removed and followed by extensive minimization to ensure that the result does not contain artificial stresses.

Stage four included a domain docking by molecular graphic. Manipulation by the rotation of selected bond dihedral angles was carried out on an Evans and Sutherland Picture System. The bonds which enhanced docking tended to be those in a bridging loop, but bonds within the two domains were varied if this appeared to give better compact structures overall. At this stage, structures were subjected to refinement which relaxed bad van der Waal's contacts though electrostatic and hydrophobic contributions were not taken into account. Two structures formed easily in the graphic manipulations suggesting the most direct folding pathway. These structures were extensively minimized using the program and parameters of stages two and three, in order to establish a conformation in the deepest accessible energy minimum. Some thousands of minor local minima were negotiated in this procedure.

One structure blew open early in minimization to a conformation over 100 kcal/mol^{-1} greater and was rejected as improbable Another of the lowest energy is shown in Figure 3. This structure has the merit to allow assignment of the most likely antigenic regions as the most accessible one. The agreements with experimentally determined antigenic determinants were quite good although the authors proposed several plausible structures differing by 10 to 15 kcal and their disulfur bridges.

All these methods still cannot report an accurate fold based on sequence alone for a reasonably sized protein. However, many applications of the same kind of software can be envisaged in different directions such as peptide drug design, energy refinement of crystallographic structure, modeling of proteins using the known conformations of homologous proteins, as well as approximate models of the structure of protein products of oncogenes for applications in artificial vaccine design.

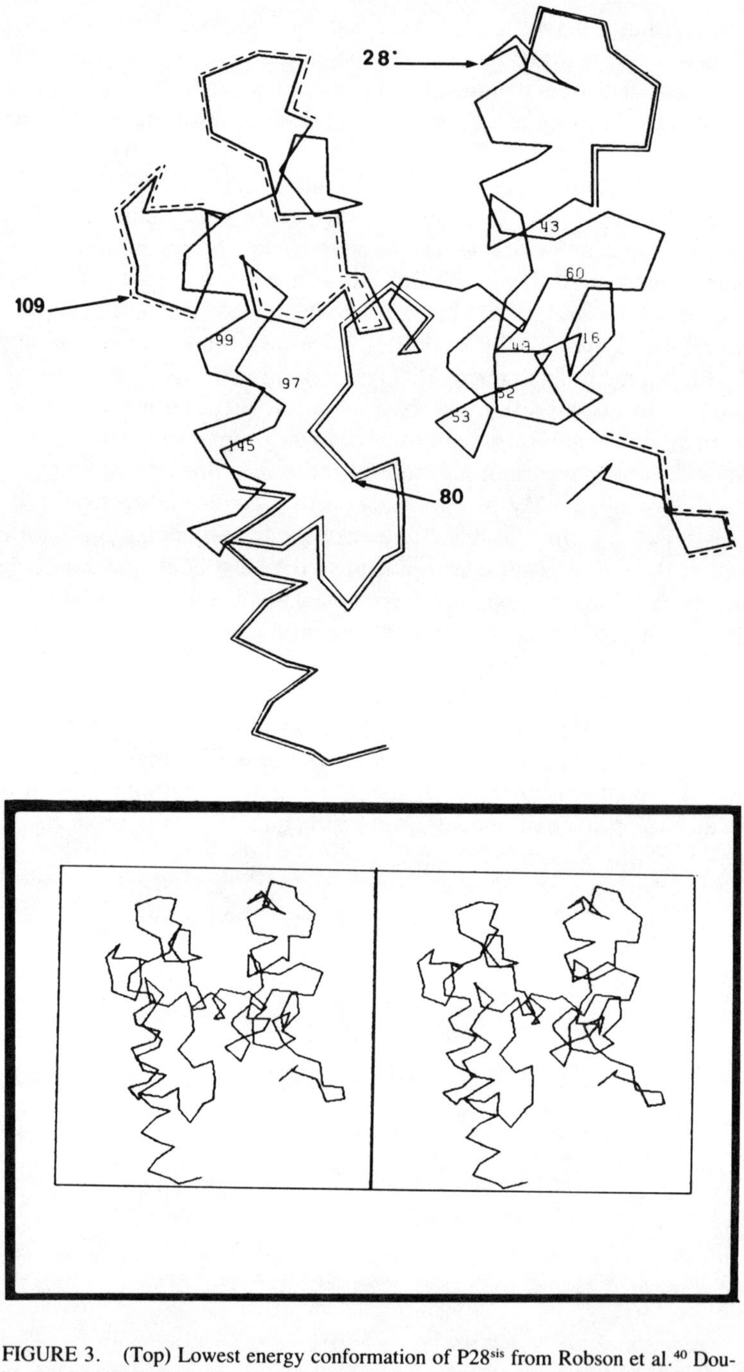

FIGURE 3. (Top) Lowest energy conformation of P28[sis] from Robson et al.[40] Double-lined backbone joining the C_α corresponds to known antigenic sites, the dotted backbone corresponds to not yet tested antigenic sites. Arrows indicate points of enzymatic cleavage. Numbers correspond to cysteine residues. (Bottom) A stereo picture of the backbone C_α. The two lobes are clearly visible on this computer drawing and the loop 61-90 is well exposed to the solvent. A structure of slightly higher energy which equally well predicts the antigenic sites has also been published and may be more consistent with the limited experimental data. (From Garnier, J., *J. Chim. Phys.*, 82, 591, 1985. With permission.)

REFERENCES

1. **Doolittle, R. F., Hunkapiller, M. W., Hood, L. E., Devare, S. G., Robbins, K. C., Aaronson, S. A., and Antoniades, H. N.,** *Science,* 221, 275, 1983.
2. **Waterfield, M. D., Scrace, G. T., Whittle, N., Stroobant, P., Johnsson, A., Wasteson, A., Westermark, B., Heldin, C. H., Huang, J. S., and Deuel, T. F.,** *Nature (London),* 304, 35, 1983.
3. **Yamamoto, T., Nishida, T., Miyaijima, N., Kawai, S., Ooi, T., and Toyoshima, K.,** *Cell,* 35, 71, 1983.
4. **Downward, J., Yarden, Y., Mayes, E., Scrace, C., Totty, N., Stockwell, P., Ullrich, A., Schlessinger, J., and Waterfield, M. D.,** *Nature (London),* 307, 521, 1984.
5. **Beug, H. and Hayman, M. Y.,** *Cell,* 36, 963, 1984.
6. **Schmidt, J. A., Beug, H., and Hayman, M. J.,** *EMBO J.,* 4, 305, 1985.
7. **Vennstrom, B., Weizsacker, F. V., Frykberg, L., and Jansson, M.,** *Eur. Mol. Biol. Lab. Rep.,* 1985.
8. **McLachlan, A. D.,** *J. Mol. Biol.,* 61, 409, 1971.
9. **Ullrich, A., Coussens, L., Hayflick, J. S., Dull, T. J., Gray, A., Tam, A. W., Lee, J., Yarden, Y., Libermann, T. A., Schessinger, J., Downward, J., Mayes, E. L. V., Whittle, N., Waterfield, M., and Seeburg, P. H.,** *Nature (London),* 309, 418, 1984.
10. **Privalsky, M. L., Ralston, R., and Bishop, J. M.,** *Proc. Natl. Acad. Sci. U.S.A.,* 81, 704, 1984.
11. **Barker, W. C. and Dayhoff, M. O.,** *Proc. Natl. Acad. Sci. U.S.A.,* 79, 2836, 1982.
12. **Lorincz, A. T. and Reed, S. I.,** *Nature (London),* 307, 182, 1984.
13. **Van Beveren, C., Galleshaw, J. A., Jonas, V., Berns, A. J. M., Doolittle, R. F., Donoghue, D. J., and Verma, I. M.,** *Nature (London),* 289, 258, 1981.
14. **Hampe, A., Gobet, M., Sherr, C. J., and Galibert, F.,** *Proc. Natl. Acad. Sci. U.S.A.,* 81, 85, 1984.
15. **Kitamura, N., Kitamura, A., Toyoshima, K., Mirayama, Y., and Yoshida, M.,** *Nature (London),* 297, 205, 1982.
16. **Reddy, E. P., Smith, M. J., and Srinivasan, A.,** *Proc. Natl. Acad. Sci. U.S.A.,* 80, 3623, 1983.
17. **Gay, N. J. and Walker, J. E.,** *Nature (London),* 301, 262, 1983.
18. **Wierenga, R. K. and Hol, W. G. S.,** *Nature (London),* 302, 842, 1983.
19. **Leberman, R. and Enger, U.,** *EMBO J.,* 3, 339, 1984.
20. **Lochrie, M. A., Hurley, J. B., and Simon, M. I.,** *Science,* 228, 96, 1985.
21. **Jurnak, F.,** *Science,* 230, 32, 1985.
22. **Halliday, K. R. J.,** *Cyclic Nucleotide Protein Phosphorylation Res.,* 9, 435, 1984.
23. **Laporte, D. C.,** *TIBS,* 10, 466, 1985.
24. **Weinberger, C., Hollenberg, S. M., Rosenfeld, M. G., and Evans, R. M.,** *Nature (London),* 318, 670, 1985.
25. **Pain, R. and Robson, B.,** *Nature (London),* 227, 62, 1970.
26. **Robson, B. and Pain, R.,** *J. Mol. Biol.,* 58, 237, 1971.
27. **Robson, B.,** *Biochem. J.,* 141, 853, 1974.
28. **Chou, P. Y. and Fasman, G. D.,** *Biochemistry,* 13, 211, 1974.
29. **Garnier, J., Osguthorpe, D. J., and Robson, B.,** *J. Mol. Biol.,* 120, 97, 1978.
30. **Richardson, J. S.,** *Adv. Protein Chem.,* 76, 241, 1973.
31. **Go, M.,** *Nature (London),* 291, 90, 1981.
32. **Go, M.,** *Adv. Biophys.,* 19, 91, 1985.
33. **Gilbert, W.,** *Nature (London),* 271, 501, 1978.
34. **Blake, C. C. F.,** *Nature (London),* 273, 267, 1978.
35. **Sternberg, M. J. E. and Taylor, W. R.,** *FEBS Lett.,* 175, 387, 1984.
36. **McCormick, F. M., Clark, B. F., La Cour, T. F. M., Kheldgaard, N., Norskov Lauretsen, L., and Nyborg, J.,** *Science,* 230, 78, 1985.
37. **Seeberg, P. H., Colby, W. W., Capon, D. J., Goeddel, D. V., and Levinson, A. D.,** *Nature (London),* 312, 71, 1984.
38. **Robson, B. and Garnier, J.,** *Introduction to Proteins and Protein Engineering,* Elsevier, Amsterdam, 1986.
39. **Pincus, M. R., Van Renswoude, J., Harford, J. B., Chang, E. H., Carty, R. P., and Klausner, R. D.,** *Proc. Natl. Acad. Sci. U.S.A.,* 80, 5253, 1983.
40. **Robson, B., Platt, E., Finn, P. W., Millard, P., Gibrat, J. F., and Garnier, J.,** *Int. J. Pept. Protein Res.,* 25, 1, 1985.
41. **Nemethy, G. and Scheraga, H. A.,** *Annu. Rev. Biophys.,* 10, 239, 1977.
42. **Kabsh, W. and Sander, C.,** *FEBS Lett.,* 155, 179, 1983.
43. **Hagler, A. T., Huler, E., and Lifson, S.,** *J. Am. Chem. Soc.,* 96, 5319, 1974.
44. **Hagler, A. T. and Lifson, S.,** *J. Am. Chem. Soc.,* 96, 5327, 1974.
45. **Robson, B. and Osguthorpe, D. J.,** *J. Mol. Biol.,* 132, 19, 1979.

46. **Hagler, A. T., Stern, P. S., Sharon, R., Becker, J. M., and Naider, F.,** *J. Am. Chem. Soc.*, 101, 6842, 1979.
47. **Davidon, W. C.,** *Comput. J.*, 10, 406, 1968.
48. **Fletcher, R. and Powell, M. J. D.,** *Comput. J.*, 6, 163, 1963.
49. **Abe, H., Braun, W., Noguti, T., and Go, N.,** *Comput. Dem.*, 8, 239, 1984.
50. **Nelder, J. A. and Mead, R.,** *Comput. J.*, 7, 308, 1965.
51. **Levitt, M. and Warshel, A.,** *Nature (London)*, 253, 694, 1975.
52. **Robson, B. and Platt, E.,** *J. Mol. Biol.*, 188, 259, 1986.

Chapter 4

MOLECULAR CONFORMATION AND POLYCYCLIC AROMATIC HYDROCARBON (PAH) CARCINOGENESIS

B. D. Silverman

TABLE OF CONTENTS

I. INTRODUCTION

The great range of carcinogenic activity induced by different PAH, Figure 1, has been one motivating factor in attempts at ranking this activity with a number or a set of numbers obtained from calculation.[1-3] For example, the pentacyclic PAH, benzo(*a*)pyrene (BP), molecule (h) of Figure 1, is a strong carcinogen whereas benzo(*e*)pyrene (BE), molecule (i) of Figure 1, having the same number of rings fused together in a somewhat different fashion is relatively inactive.[4] Furthermore, methylating the relatively weak carcinogen, benz(*a*)anthracene (BA), molecule (d) of Figure 1, at the 7 and 12 positions is known to yield one of the most highly active PAH carcinogens, namely, 7,12-dimethylbenzanthracene (DMBA), Figure 2. Such dramatic changes in carcinogenic potency upon geometric rearrangement of the fused six-membered rings of PAH or by the addition of substituents has intrigued investigators both past and present.

Since all of the early calculations were performed on parent molecules during a period of time when the route of activation and the importance of PAH metabolism was poorly understood, correspondence between the results of these early calculations and the relative carcinogenic strength of different PAH might be considered fortuitous.[1,5,6]

In fact, after appreciating the complexity of the process of carcinogenic initiation, one might wonder how the computation of a small set of numbers can contribute at all to our understanding of how PAH can induce carcinogenesis! One answer to this is that over the past decade or more a number of developments have occurred that have significantly narrowed our focus on the chemical events of prime importance in triggering the initial stages of PAH carcinogenesis. These developments include

1. Recognition of the correspondence between the ability of a PAH to bind covalently to DNA in vivo and its ability to induce tumorigenesis.[7,8]
2. Identification of the 7,8-diol-9,10-epoxide of BP, Figure 3, as the ultimate carcinogen of BP, or in other words, as that particular metabolic molecular species that chemically interacts with DNA and is apparently responsible for BP-induced tumorigenesis.[9,10]
3. Recognition of the importance of diol-epoxides generally in PAH-induced carcinogenesis.[11,12]
4. The generalization and subsequent experimental verification that for, perhaps most, if not all odd alternant PAH (PAH consisting of fused six-membered rings only) the bay-region diol-epoxides are the most tumorigenic of all the potential metabolites.[12-14]

The picture that has emerged concerning the route of activation of PAH to chemical species that cause cancer is schematically illustrated in Figure 4 with use of BP as an example. First (Step a), the parent molecule is metabolized to an epoxide by the ubiquitous oxygenating enzymes, the cytochrome $P_{450's}$. This epoxide is then ring-opened metabolically by the epoxide hydrolases (Step b) to yield the so-called "proximate carcinogen", the 7,8-diol of BP. Next (Step c), a second epoxidation step, in the bay-region, induced by the cytochrome $P_{450's}$ yields the bay-region diol-epoxide, the 7,8-diol-9,10-epoxide of BP. This "ultimate carcinogen", or chemical species believed to interact directly with DNA, is believed to spontaneously ring open (Step d) yielding a reactive carbocation which interacts with DNA (Step e). A preferential site of interaction is the N2-exocyclic amino group of the base, guanine[15,16] as shown (Step e). Such interaction compromises the genetic fidelity of DNA and is apparently a first step along the path to carcinogenesis.

Concurrent with the development of this picture, Jerina and co-workers[13,17,18] had pointed out that calculated delocalization energies of carbocations formed by epoxide ring opening of bay-region diol-epoxides can yield a measure, albeit imperfect, of the carcinogenic activity induced by different PAH. Calculated carbocation delocalization energies provide a relative

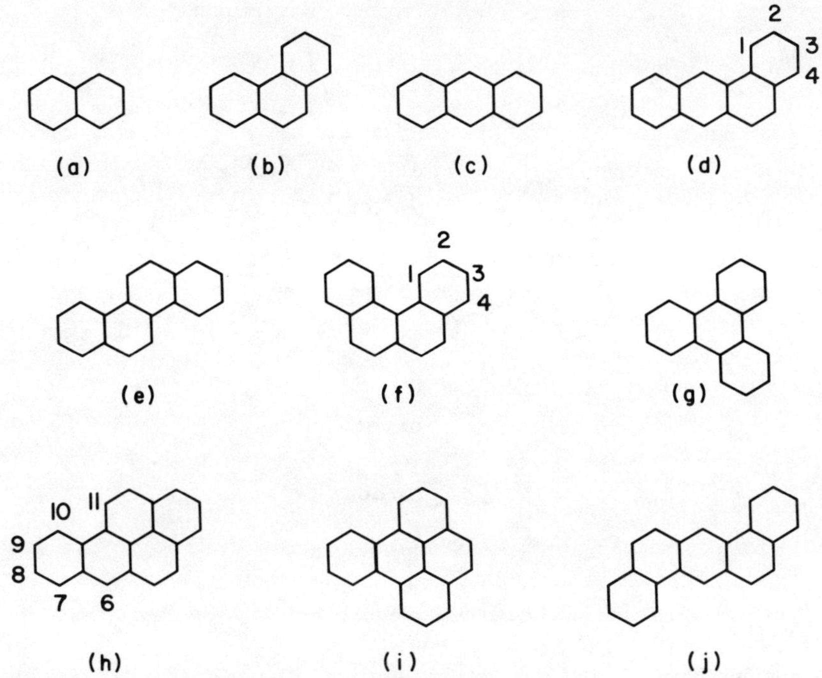

FIGURE 1. Polycyclic aromatic hydrocarbons.

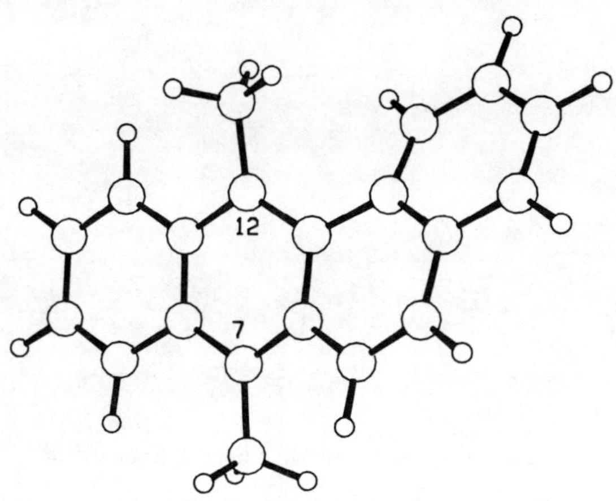

FIGURE 2. 7,12-Dimethylbenz(*a*)anthracene (DMBA).

measure of the ease of spontaneous epoxide ring opening, (Step d of Figure 4), and hence a relative measure of one aspect of diol-epoxide chemical reactivity. One might expect this aspect of diol-epoxide reactivity to be important in determining the relative carcinogenic potency of different PAH if indeed diol-epoxide ring opening is a rate-limiting step.

Calculations by Jerina and co-workers have shown:[13,14]

1. That the carbocation formed by ring opening of the bay-region diol-epoxide of PAH has a greater delocalization energy associated with it than other carbocations formed by ring opening of any of the other isomeric diol-epoxides, e.g., in the case of BP, the delocalization energy of the carbocation obtained by ring opening of the bay-region

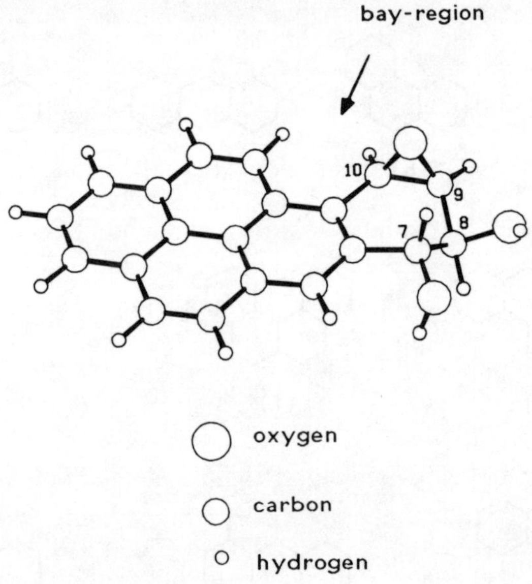

FIGURE 3. 7,8-Diol-9,10-epoxide of benzo(*a*)pyrene (BP).

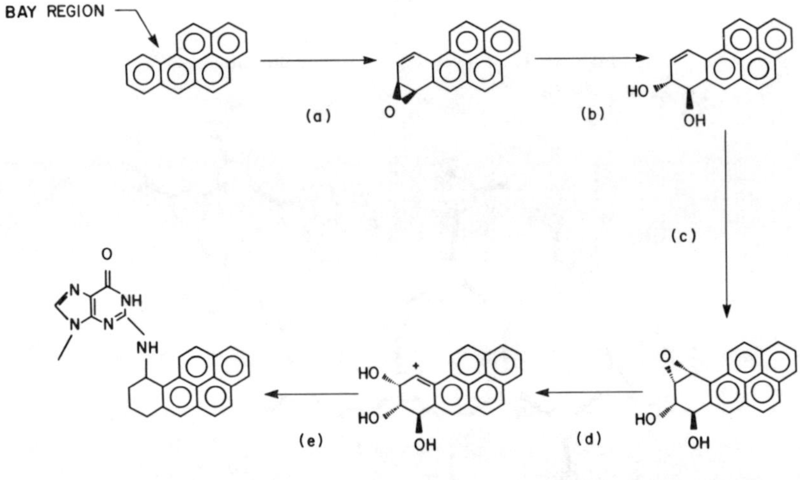

FIGURE 4. Route of BP activation to an ultimate carcinogen.

diol-epoxide, (Step d of Figure 4), is greater than that of the carbocation obtained from epoxide ring opening of the 9,10-diol-7,8-epoxide of BP.

2. That the relative magnitudes of the bay-region carbocation delocalization energies do provide a reasonable relative measure of a PAHs ability to induce carcinogenesis.

These results suggest that ease of epoxide ring opening is an important rate-limiting step in initiating the carcinogenic process induced by PAH and that it is the bay-region diol-epoxide that is the ''ultimate carcinogen''.

It is now known, however, that besides ease of epoxide ring opening, there are a number of other topological or stereochemical features of PAH and PAH metabolites that are important in determining the ability of a PAH to induce carcinogenesis. For example, the bay-region diol-epoxides of BP are found as one of two diastereomers, namely, (±)-7α,8β-dihydroxy-9β,10β-epoxy-7,8,9,10-tetrahydrobenzo(*a*)pyrene (anti-BPDE), Figure 3 and (±)-

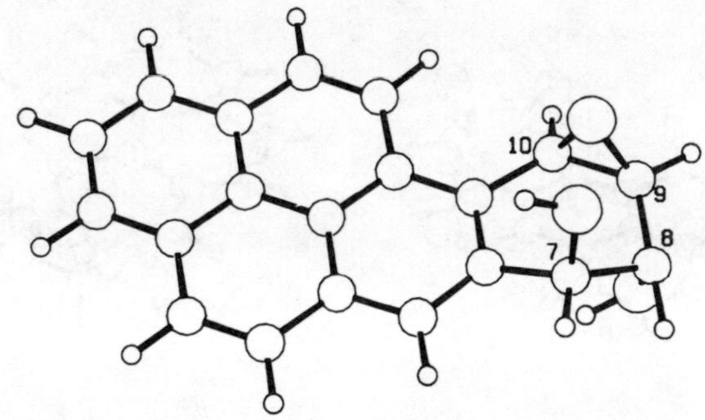

FIGURE 5. 7α,8β-Dihydroxy-9α,10α-epoxy-7,8,9,10-tetrahydrobenzo(*a*)pyrene (syn BPDE).

7α,8β-dihydroxy-9α,10α-epoxy-7,8,9,10-tetrahydrobenzo(*a*)pyrene (syn-BPDE), Figure 5. Anti and syn refer to whether the epoxide and benzylic hydroxyl groups (the hydroxyl group adjacent to the aromatic portion of the molecule) are on the opposite or on the same side of a plane through the aromatic part of the molecule. Anti-BPDE has been found to be more markedly mutagenic and tumorigenic than syn-BPDE in mammalian systems. Consistent with this, the anti-diastereomer of BP has also been found to bind to a greater extent to rat embryo DNA than the syn-diastereomer.[19] One apparent difference between the two diastereomers is the orientation of the hydroxyl groups with respect to the aromatic part of the molecule. The anti-BPDE diastereomer, Figure 3, has hydroxyl groups that are quasi-diequatorial with respect to the aromatic plane of the molecule, whereas the syn-diastereomer has hydroxyl groups that are quasi-diaxial. One might surmise that this conformational difference could play a role in the ability of a PAH diol-epoxide to bind to DNA. Perhaps the bulkier quasi-diaxial conformation of the hydroxyl groups inhibits the initial molecular intercalation[20] prior to the PAHs covalent interaction with DNA.

There is one additional topological degree of freedom. Each of the two diastereomers can, in principle, exist in one of two different stable conformational states.[21] The conformers generally observed, are the ones we have discussed in connection with BP, Figures 3 and 5. These have similar epoxide and angular benzo-ring geometry. For these conformers, the plane of the epoxide ring is approximately perpendicular to the aromatic plane of the molecule. Also, the geometry of the angular benzylic ring is such that certain carbon-hydrogen bonds on this ring adopt the ethylenic staggered as opposed to the less energetically favored eclipsed geometry.

The two conformers that are not generally observed in solution, namely the anti-diaxial, Figure 6a, and syn-diequatorial, Figure 6b, also have epoxide and angular benzo-ring geometry similar to each other. For these conformers the plane of the epoxide is tilted away from the normal to the aromatic plane and certain carbon-hydrogen bonds on the metabolized benzo-ring adopt a less energetically favorable eclipsed geometry. It is of interest that the syn-diequatorial conformer has recently been observed in a sample of crystalline BP diol-epoxide.[22]

For the PAH, benzo(*c*)phenanthrene (BCP), one of these conformers, the syn-diequatorial diol-epoxide, has actually been observed in solution.[23] In an interesting series of experiments,[14,23-26] Jerina and co-workers have observed a syn-diequatorial diastereomer of the bay- or "fjord-region" diol-epoxide of benzo(*c*)phenanthrene. BCP is a tetracyclic PAH, Figure 7, with a fjord-region between its two terminal rings. There is significant steric crowding in this region and the molecule, therefore, deviates significantly from planarity.[27,28] If the molecule was flattened onto a plane, the two hydrogen atoms in the fjord-region would be approximately half an angstrom apart. It is the steric crowding between these two hydrogen

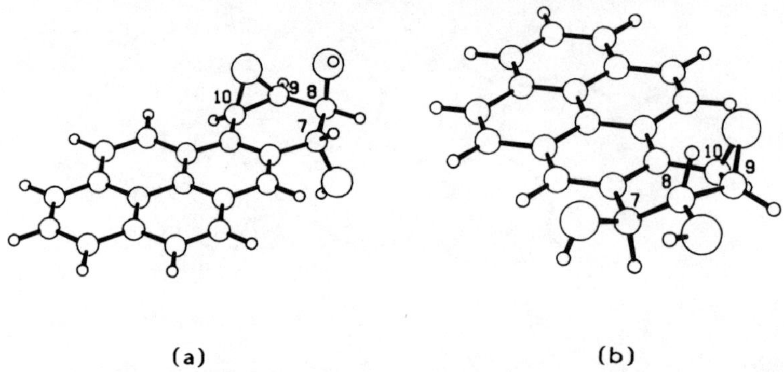

FIGURE 6. (a) Anti-diaxial diol-epoxide of BP and (b) syn-diequatorial diol-epoxide of BP.

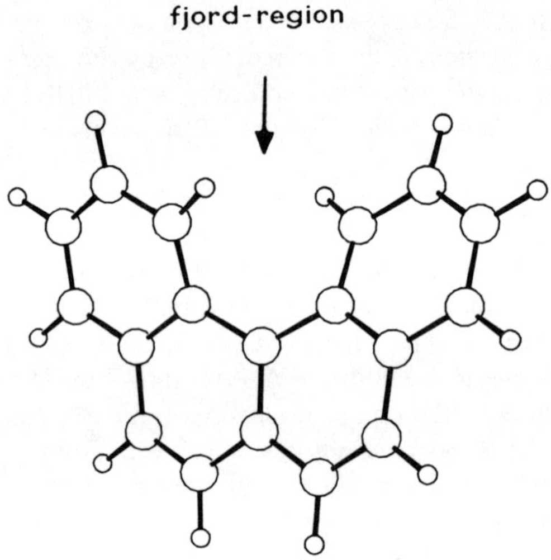

FIGURE 7. Benzo(*c*)phenanthrene (BCP).

atoms that is responsible for the significant deviation from planarity of BCP. Steric crowding is also present for the fjord-region diol-epoxides and it will be shown that calculation indicates that the syn-diequatorial diol-epoxide, Figure 8c, should be more stable than the syn-diaxial conformer, Figure 8b.

It is of interest that both syn-diequatorial, Figure 8c, and anti-diequatorial, Figure 8a, fjord-region diol-epoxides of BCP, have been shown to exhibit similar activity on mouse skin.[24] This is the first example of a syn-diastereomer exhibiting comparable tumorigenic activity to the anti-diastereomer. It implicates hydroxyl group orientation as an important determinant of tumorigenic activity.

The present chapter will be devoted to an examination of the calculated effect of steric crowding on PAH molecular conformation and consequently upon PAH reactivity. First we will review the calculations performed on the sterically crowded diol-epoxides of BCP.[27,28] Molecular structures obtained from calculation will be illustrated to highlight certain of the differences in geometry that are apparently responsible for the large differences observed in tumorigenic activity. The calculated relative stabilities of the different fjord-region diol-epoxides will be examined.

Next, we will discuss the results of calculation of the relative stability of the various bay-

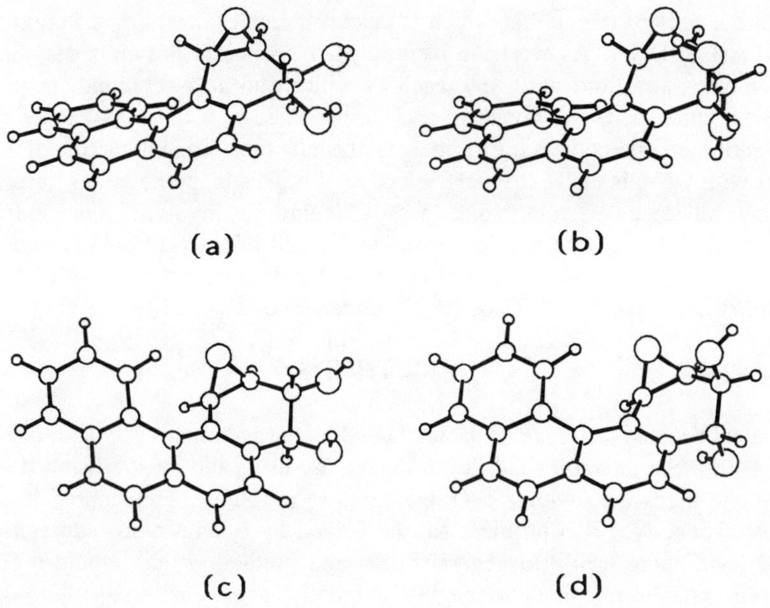

FIGURE 8. (a) Anti-diequatorial diol-epoxide of BCP; (b) syn-diaxial diol-epoxide of BCP; (c) syn-diequatorial diol-epoxide of BCP; and (d) anti-diaxial diol-epoxide of BCP.

region diol-epoxides of DMBA. It has been generally observed that methylation in the bay-region of PAH generally increases the ability of a PAH to initiate carcinogenesis.[29-32] One question that we will address is whether the addition of a methyl group in the bay-region, such as found in DMBA, destabilizes conformers of the DMBA diol-epoxide diastereomers in a manner similar to that found for the sterically crowded diol-epoxides of BCP. Recent measurements of the binding of diol-epoxide diastereomers to DNA indicate significant binding of the syn-isomer.[33] Perhaps the enhanced carcinogenic activity of DMBA is at least partially related to hydroxyl group orientation influenced by the presence of the methyl group in the bay-region.

Finally, we will examine the effects of steric crowding due to methylation in the bay-region of chrysene and 15,16-dihydrocyclopenta(*a*)phenanthren-17-one (CPP). For these PAH as well as for DMBA we find that steric crowding destabilizes the conformers that are generally observed, namely the anti-diequatorial and syn-diaxial diastereomers with respect to the conformers not commonly observed, namely, the anti-diaxial and syn-diequatorial diastereomers. Actually, an anti-diaxial diol-epoxide conformer resulting from in vivo or in vitro metabolism of any PAH has yet to be observed.

We have also calculated the relative bay-region carbocation delocalization energies. This provides a measure of the relative ease of epoxide ring opening for the monomethyl isomers that do not have a methyl group on the potentially active benzo-ring. The results of these calculations indicate that the direct or ''through-space'' interaction between the carbocation center and the methyl group enhances the carbocation stability of the bay-region methylated isomer with respect to the other isomers. This is seen simply since Simple Huckel calculations, which treat only ''through-bond'' interactions, and neglect ''through-space'' interactions, yield a low value of carbocation delocalization energy for the bay-region methylated isomer compared with the other methylated isomers. Delocalization energy results will be given for the methylated chrysenes and cyclopentaphenanthrenes.

Finally, we should mention one other topological degree of freedom that will not be examined in the present paper, but which is of considerable interest and importance in the binding of PAH to DNA. As we have mentioned, bay-region diol-epoxides can be found as diastereomeric pairs (anti and syn), each of which can also be found, in principle, as conformational pairs (perp and nonperp). Furthermore, each of these molecules can also be found as enantiomeric pairs or as mirror images of each other. The interaction of the different enantiomers with DNA is also extremely selective.[34,35] This is, perhaps, not surprising since DNA is chiral, having a particular handedness. Calculations, involving molecular modeling, have shown why different diol-epoxide enantiomers will interact differently with DNA.[36,37] Why the (+) diol-epoxide enantiomer interacts so much more strongly with DNA than the (−) enantiomer is, however, still not clearly understood.

II. METHODS

As computers become faster and capable of handling larger amounts of data, one is tempted to keep up with these technological developments by using the most advanced state of the art software and hardware to solve even the simplest problems. This approach is generally not the most productive. If a problem can be solved by a simple procedure, the solution should lend itself more easily to generalization and prediction than would a complicated procedure. This enables one to avoid lengthy, detailed, and expensive calculations. So called *ab initio* programs which calculate molecular ground state properties from scratch, with relatively few assumptions or approximations, require computer processing times (CPU) on the order of hours for the molecules of interest to us.

Therefore, one factor that has motivated our computational research on carcinogenic PAH has been the desire to distill easily applicable rules obtained from Simple Huckel theory concerning ease of epoxide ring opening or chemical reactivity. Much of this work has been presented in a recent review[3] and will be treated by a chapter in another volume of the CRC Press Uniscience Series. Such simple rules should enable experimentalists to obtain quick ''rule of thumb'' estimates of relative chemical reactivity for PAH diol-epoxides of interest.

Of particular interest are the simple rules concerning the effect of methylation at so-called ''remote substitutional sites'' of PAH.[38,39] ''Remote substitutional sites'' are sites that are neither on the potentially reactive benzo-ring nor at the two positions that are adjacent to this ring, namely the bay-region position and the peri position.

The electronic effects of methylation at ''remote substitutional sites'' appear to parallel the induced relative differences in carcinogenic activity for a number of different PAHs.[38-41] Apparently for these ''remote substitutional sites'', induced differences in metabolism arising from methylation are not extreme[42-44] and the induced changes in molecular reactivity (ease of epoxide ring opening) appear to be an important determinant of relative carcinogenic activity.

For methylation at ''nonremote positions'', the rules inferred from Simple Huckel theory may be either incorrect or inappropriate. For example, methylation at the peri position of PAH, at position 6 of BP, molecule (h) of Figure 1, a position which we have defined as nonremote is known[45] to not only inhibit metabolism to the proximate 7,8-diol of BP, but also to inhibit metabolism to the ultimate carcinogen of BP, the 7,8-diol-9,10,-epoxide (Figure 3). Hence, no matter how methylation at position 6 might affect bay-region diol-epoxide reactivity through electronic effects, such effects might be irrelevant in ranking the carcinogenic activity of this isomer. Studies of BP as well as of other PAH lead one to suspect that peri methylation of PAH might generally inhibit metabolism to the important reactive intermediates.[45,46]

Other substitutional positions for which induced changes of the ease of epoxide ring opening upon methylation appear to be unimportant are the positions on the reactive benzo-ring. In fact, the inhibition of ring oxygenation and consequent loss of carcinogenic activity upon such methylation was initially used as evidence supporting identification of bay-region diol-epoxides as important reactive intermediates.[18]

The effect of methylation in the bay-region of a PAH, or in other words, at the bay-region position adjacent to the angular benzo-ring, for example, at position 11 of BP, molecule (h) of Figure 1, is of special interest. Methylation at this position, a ''nonremote position'' by definition, has been found to enhance the tumorigenicity induced by a number of PAH.[29-32] In fact, aside from BA, methylation in the bay-region has yielded the most tumorigenic isomer of all potential monomethyl isomers of a particular PAH. For BA, the bay-region methylated isomer, 12-methylbenzanthracene (12-MBA) is less active than 7-MBA (e.g., Reference 1).

In ranking the ease of epoxide ring opening for bay-region diol-epoxide methylated isomers, Simple Huckel[47] and Extended[48] Huckel theory yield totally different results. One infers from Simple Huckel theory that the bay-region methylated iosmer should be one of the least reactive isomers, whereas Extended Huckel theory indicates this isomer should be one of the most reactive isomers. This difference can be traced to the neglect of the direct, ''through-space'', interaction between the methyl group and carbocation center for the Simple Huckel calculation. Simple Huckel theory neglects this interaction since it treats interactions between only neighboring atoms of the molecule that are covalently coupled, namely, only the ''through-bond'' interactions. The Extended Huckel calculation,[48] however, treats interactions between all atoms of the molecule. The proximity of the methyl group to the carbocation center, therefore, invalidates the use of Simple Huckel theory for obtaining carbocation delocalization energies of the bay-region methylated isomers of PAH. This is one reason that we have designated bay-region methylation as a ''nonremote substitution'', i.e., not to be treated by Simple Huckel theory. There is, however, one other reason that one might designate methylation in the bay-region of PAH as a ''nonremote'' substitution. Examination of the distances between the hydrogen atoms of the methyl group in the bay-region and one of the hydrogen atoms on the epoxide ring indicates that the bay-region diol-epoxide should be strained due to steric crowding. Similar steric crowding should also be present for the ring-opened carbocation as well as for the parent molecule.

The following two concerns therefore arise. First, if the molecule is deformed, how does this affect the ease of epoxide ring opening? For example, if the diol-epoxide is more strained than the ring-opened carbocation, the ease of ring opening could be enhanced by steric crowding. Second, what implications do steric crowding hold with respect to the relative stability of the various bay-region diol-epoxide conformers? Do effects of steric crowding resulting from bay-region methylation parallel the effects of steric crowding in the fjord-region of BCP previously discussed? Consideration of the effects of steric crowding upon molecular reactivity and conformational stability requires more than inferences from Simple Huckel theory.

The present paper, therefore, represents a departure from the previous work aimed at developing simple rules concerning diol-epoxide reactivity. In treating the effect of methylation at the ''nonremote'' bay-region substitutional site we will attempt to do as detailed and complete a calculation as possible.

The procedure that we have adopted to calculate the ground state energies of different molecular species is as follows. First, an ideal or starting geometry is determined by ''walking'' around the molecule using the CDNT program.[49] The ''walk'' around the molecule consists of starting at a given atom of the molecule. The full molecular geometry is then specified by designating sequentially, the bond length, bond angle, and dihedral angle of each subsequent atom. Representative values for these parameters are chosen. The CDNT program converts this ''Z-matrix'' (bond lengths, bond angles, and dihedral angles) to Cartesian coordinates. It also yields interatomic distances which can be used as a quick check on any errors made in generating the Z-matrix. The Cartesian coordinates are then used as input to the Allinger force field program.[50] The force field program is primarily a phenomenological "ball and spring" type program which uses empirically determined spring constants and equilibrium bond lengths, bond angles, and dihedral angles to define the force-field energy. The program varies the molecular geometry until a local minimum in the force field energy is found. One strength of this particular force field program for the problem at hand is that it does perform a Huckel type calculation to assist in determining subtle differences in the geometry of π aromatic structures. As a result of this procedure it is capable of yielding the small differences in bond lengths observed for parent PAH structures. For example, it yields BP and BCP structures that closely parallel the X-ray-determined structures.[21,27] Other force-field programs that are available do not have this capability. A parameter set for the different types of atoms is provided with the program, however, the parameters can also be user defined. We have supplemented the built-in program parameters with a parameter set appropriate for epoxides.[21] This, of course, will yield more representative geometries than obtained by treating the C–O–C linkage of the epoxide, approximately, i.e., as an ether linkage. For carbocation geometries, on the other hand, we have used only the built-in force-field parameters. This at least simulates bond length differences associated with differences in single and double bond character. It neglects, however, changes in geometry associated with the distribution of electronic charge about the carbocation center. Furthermore, all of the force-field calculations have been performed with the input option that turns off all electrostatic interactions between the atoms, i.e., turns off the coulomb and dipole forces. For PAH, this was found to yield structures that most closely parallel observed X-ray structures. A number of different input geometries were also chosen for each run to check whether the program had found the most stable of locally stable equilibrium geometries. The lowest energy geometry was then chosen in further calculations.

Even though one obtains a force-field energy associated with the stable force-field structure this is not always the most reliable measure of the relative stability of several closely related structures.[51] We have, therefore, used the force-field determined structures as input to molecular orbital programs for a determination of total molecular energy. Programs utilized have been those of the Gaussian series,[52] Gamess,[53] as well as semiempirical programs such as INDO[54] and Extended Huckel.[48] Whereas the absolute Hartree-Fock energies one obtains from these programs are in error by several electronvolts, the relative error for the closely related structures that we will examine is sufficiently small to provide useful information concerning the relative stability of these structures. Further details concerning the methods used will be discussed in the section detailing the results.

The sequence of presentation is not the sequence in which the calculations were performed. Since the calculations to be described were motivated in part by the results of preceding calculations as well as by experiment, a word about this might be useful. Calculations were initially performed on the epoxides of the benzo(*c*)phenanthrenes since the initial experimental results of Jerina et al.[14,23-26] were so striking. The calculations were performed essentially to see if steric crowding in the fjord-region of BCP did indeed destabilize the syn-diaxial diastereomer of the BCP diol-epoxide with respect to its syn-diequatorial conformer. This is what the calculations indicated.[27,28] The next calculations were performed for epoxides of the bay-region methylated PAH benz(*a*)anthracene (BA), cyclopenta(*a*)phenanthrene, and chrysene.[55] These calculations were motivated by the thought that steric crowding induced by methylation in the bay-region might destabilize the diol-epoxide conformers in a manner similar to that found for BCP. It was thought that this, perhaps, might partially explain the enhanced tumorigenic activity of the bay-region methylated isomers of PAH. Such destabilization was found for the methylated PAH epoxides but to a lesser extent than found for BCP. The BCP calculations were performed for diol-epoxides whereas the other calculations, for BA and CPP, were performed with neglect of the diol, namely for epoxides. For simplicity the hydroxyl groups were replaced by hydrogen

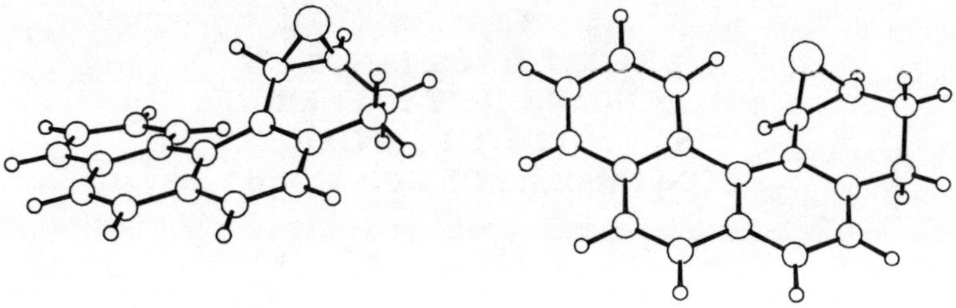

(a)

(b)

FIGURE 9. Epoxide conformers of BCP. (a) Perp and (b) nonperp.

atoms. For BCP, the orientation of the diol and its influence on relative conformer stability was treated in a relatively simple manner. Treating the orientation of the diol and its influence on molecular conformational stability was a problem that seemed initially to be of little consequence and one that did not warrant undue effort. This problem was finally tackled in a more complete but still imperfect manner, for the following reason.

The BCP and other calculations indicated an interesting parallelism with respect to the relative stability of the conformers of the diastereomers. If the syn-diaxial conformer was destabilized by steric crowding in a fjord- or bay-region, the anti-diequatorial conformer should likewise be destabilized. All in vitro evidence (primarily NMR shifts for BCP, see Reference 23), however, indicates that, whereas, the syn-diastereomer has a quasi-diequatorial conformation, the anti-diastereomer is also quasi-diequatorial and not diaxial as one might have surmised from the results of calculation. This expected parallelism of structures led Sayer et al.[23] to speculate that the loss of parallelism might be due to greater coupling between the hydroxyl groups for the diequatorial conformation compared with the diaxial conformation. This remark essentially motivated the more detailed calculations to be described for DMBA. These calculations treated the interaction between the hydroxyl groups of the diol in a more complete manner than they were treated heretofore. DMBA was chosen for calculation since binding studies[33] had shown a significant number of syn adducts. Such calculations consumed hundreds of hours of CPU time on an IBM model 3081.

The OR TEP molecular graphics program[56] has been used to generate most of the molecular structure figures.

III. RESULTS AND DISCUSSION

A. Benzo(*c*)phenanthrene (BCP)

BCP, Figure 7, and its reactive diol-epoxide metabolites, Figure 8, are sterically crowded due to the close approach of the hydrogen atoms in the fjord-region of the molecule. The implications of such crowding are determined computationally as follows. The structures of two fjord-region epoxide conformers, which for simplicity we have called the perp, Figure 9a, and nonperp, Figure 9b, have been determined with use of the Allinger force-field program. Perp refers to the conformer for which the plane of the epoxide ring is roughly perpendicular to the aromatic plane. This conformer has a benzo-ring geometry similar to that of the anti-diequatorial and syn-diaxial fjord-region diol-epoxides of BCP. In fact, in our calculations the benzo-ring geometry will be identical for these diastereomers. The epoxide geometries have been determined with a user-defined epoxide parameter set provided by Jorgensen and Snyder.[21] We have constructed the diol-epoxide geometries, Figure 8, by replacing the appropriate hydrogen atoms on the benzo-ring with hydroxyl groups while holding the force-field determined benzo-ring geometry fixed. The orientation of the hydroxyl

Table 1
TOTAL GAUSSIAN 70 MOLECULAR ORBITAL ENERGIES (STO-3G) OF THE EPOXIDE CONFORMERS OF BCP AND BA

	BCP	BA
Perp	−755.27122 (Au.)	−755.27542
Nonperp	−755.28000	−755.27300

1 Au. = 27.2 eV.

groups was determined with use of the INDO molecular orbital program,[53] i.e., the dihedral angles of the hydroxyl groups were chosen at the INDO energy minimum. We have found that this procedure actually yields lower molecular orbital energy structures than obtained by including the hydroxyl groups in the force-field structure optimization. Nonperp refers to the epoxide conformer with epoxide ring geometry that is appropriate for the anti-diaxial and syn-diequatorial diastereomers. Again, these two diastereomeric structures, Figures 8d and 8c, are obtained by adding the hydroxyl groups at the appropriate positions on the benzo-ring.

Gaussian 70 molecular orbital energies[57] have then been determined with use of the force-field epoxide geometries augmented by the diol or the hydroxyl groups. Table 1 lists the Gaussian 70 energies (STO-3G basis) for the epoxide conformers. It is interesting to note that the perp epoxide conformer of BCP, Figure 9a, is less stable than the nonperp conformer, Figure 9b, by approximately 5 kcal/mol. When such calculation is performed for the epoxides of a PAH with a nonsterically crowded bay-region, e.g., BA[28,55] the opposite result is found. Namely, the perp epoxide conformer is found to be more stable. The greater stability of the perp conformer compared with the nonperp conformer in the absence of a sterically crowded bay-region results since certain bonds on the benzo-ring adopt the more stable ethylenic type staggered conformation as opposed to the less energetically favorable eclipsed conformation. For BCP, the steric crowding involving the two hydrogen atoms in the fjord-region of the perp conformer is sufficiently great to offset the energetically unfavorable ring geometry. The corresponding diol-epoxide geometries are generated by replacing the appropriate hydrogen atoms on the benzo ring with hydroxyl groups. The relevant C–H bonds are extended to yield C–O bond lengths and hydrogen atoms are coupled to the oxygen atoms to form the two O–H hydroxyl groups. The hydroxyl group orientation (C–C–O–H dihedral angle) is obtained with use of the semiempirical INDO molecular orbital calculation. This orientation is obtained by varying only the C–C–O–H dihedral angles while keeping all other atoms fixed in the determination of the total molecular energy minimum. The final geometry with optimized hydroxyl group orientation has then been used as input to the Gaussian 70 molecular orbital program and the results given in Table 2.

For comparison, results are also given for the diol-epoxides of benzanthracene, a PAH with essentially no steric crowding in the bay-region, molecule (d) of Figure 1. For BA, examination of the table shows that the generally observed conformers, the syn-diaxial and anti-diequatorial are calculated to have greater stability then the syn-diequatorial and anti-diaxial conformers, as one might expect. Such result is, however reversed for the diol-epoxides of BCP. For BCP, the syn-diequatorial and anti-diaxial conformers are calculated to have greater stability. Comparing the values given in Tables 1 and 2, one should surmise that this difference in diol-epoxide stability is nothing more than a reflection of epoxide stability.

The present calculations, therefore, provide a basis for understanding the origin of the

Table 2
TOTAL GAUSSIAN 70 MOLECULAR ORBITAL ENERGIES (STO-3G) OF THE DIFFERENT 1,2-DIOL-3,4-EPOXIDES OF BCP AND BA

	BCP	BA
Syn-diaxial	−902.91091 (Au.)	−902.91707
Anti-diequatorial	−902.90900	−902.91555
Syn-diequatorial	−902.91801	−902.91491
Anti-diaxial	−902.91727	−902.91279

1 Au. = 27.2 eV.

observed stability of the syn-diequatorial diol-epoxide conformer of BCP. What was, however, initially surprising was the lack of any indication of the presence of an anti-diaxial conformer among the synthesized anti-diastereomers.[23]

B. 7,12-Dimethylbenz(*a*)anthracene (DMBA)

DMBA is one of the most active PAH carcinogens and as a result has been extensively studied. Since the identification of BPDE as the important BP reactive intermediate, several studies have implicated a DMBA bay-region diol-epoxide as the important reactive intermediate responsible for initiating DMBA-induced tumorigenesis.[58] More recent studies have shown that the syn bay-region diol-epoxide diastereomer binds to DNA as extensively as the anti-diastereomer.[33] This contrasts with investigations of the adducts of BPDE and DNA.[19] Since DMBA is sterically crowded, deviating significantly from planarity[59] there has been speculation concerning the relationship between its strained molecular geometry and the initiating potency of the molecule.

As previously mentioned, the question of interest to us is, whether one expects any relative change in bay-region diol-epoxide stability among the different possible conformers. As pointed out, these calculations were motivated by the thought that the enhanced binding of the syn-diastereomer to DNA might result from a quasi-diequatorial conformation of the hydroxyl groups.

The calculations were performed in a somewhat different manner from those previously described for BCP. Instead of using the semiempirical INDO calculation to determine the orientation of the two hydroxyl groups, we used the *ab initio* Gaussian 80 program.[60] This treats interactions between the two polar hydroxyl groups in a less approximate manner than the INDO program.

The computational procedure that was adopted was as follows. First, a force-field determined geometry was found for the two bay-region epoxide conformers, the perp and nonperp conformers (Figure 10). This geometry was used in constructing the various diol-epoxide geometries. At first, hydroxyl groups were appended to the appropriate molecular positions. Standard bond lengths were chosen for the C–O and O–H bonds and a standard bond angle for the C–O–H angle. We would have liked to determine the optimum (minimum energy) dihedral angle orientation of the two hydroxyl groups with an *ab initio* molecular orbital program such as Gamess[53] or one of the Gaussian series programs.[52] For example, one can optimize any input parameter of the Z-matrix used as input to the Gaussian 80 program[60] while holding all other parameters fixed. This wasn't at all possible, however, for the molecules of interest to us for two reasons. First, the integrals calculated and used in the Gaussian series programs are packed in such a manner that the program would only allow the use of up to and including 127 orbitals in the molecular orbital basis. For the STO-3G

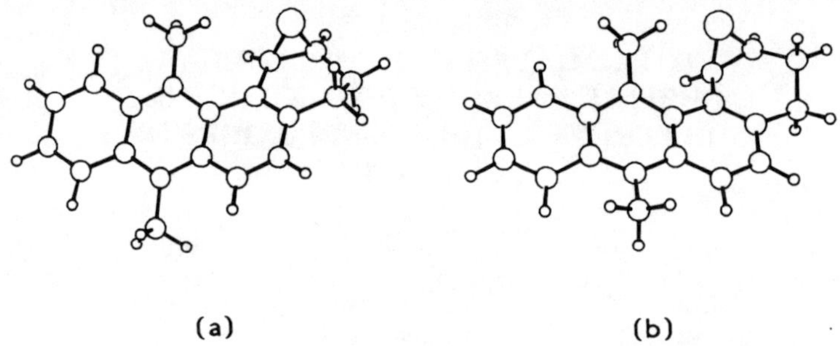

FIGURE 10. Epoxide conformers of DMBA. (a) Perp and (b) nonperp.

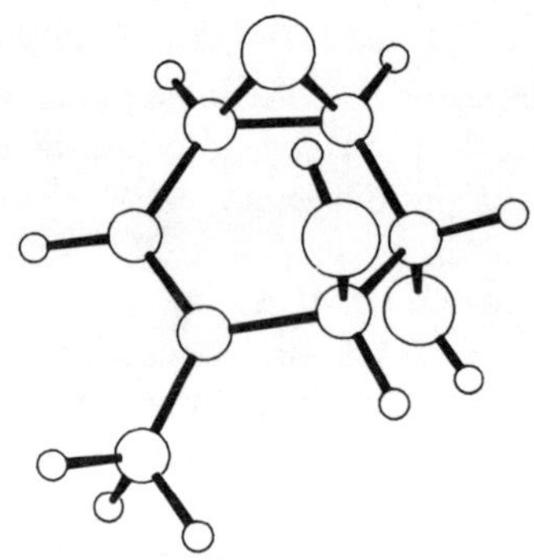

FIGURE 11. Benzo-ring fragment of a syn-diaxial diol-epoxide diastereomer.

minimum orbital basis, 133 molecular orbital functions are required. This was no such limitation with the Gamess[53] program, however, the version of the program that we had did not allow selective optimization of any set of Z-matrix parameters. Even granting this, the CPU time required to find the energy minimum for the dual hydroxyl group orientation would have been prohibitive.

We therefore adopted the following strategy to find the lowest energy orientation for the two hydroxyl groups. The molecule was truncated so that only the benzo-ring was retained with use of the fixed force-field determined geometry for the entire structure. The peri-position carbon and hydrogen atom were also retained and this peri group fragment converted to a methyl group.[61] We assumed that the relative orientation of the two hydroxyl groups forming the diol was, to a large extent, determined by the electronic and steric factors arising from the benzo-ring and peri hydrogen atom, respectively. One of the fragments that we treated is shown in Figure 11. This geometry was then used as input to Gaussian 80 molecular orbital program. The program varied the two dihedral angles of the hydroxyl groups in its search for a local energy minimum.

To avoid getting trapped in local minima, the relative hydroxyl group orientation was calculated for 16 different initial orientations. These were chosen as all possible combinations of the orientations of 0, 90, 180, and 270 degrees for the dihedral angle of each hydroxyl group. A number of different local minima were found but we believe that we have found the global minimum in this two-dimensional space of rotations since many of the different initial starting parameters converged to the same values. On the average, each initial set of dihedral angles required approximately 120 min for convergence. Since there were 16 initial sets of dihedral angle values and the program was run for the two conformers of each diastereomer, the approximate CPU time to find these energy optimized orientations was 7680 CPU minutes or about 128 CPU hours.

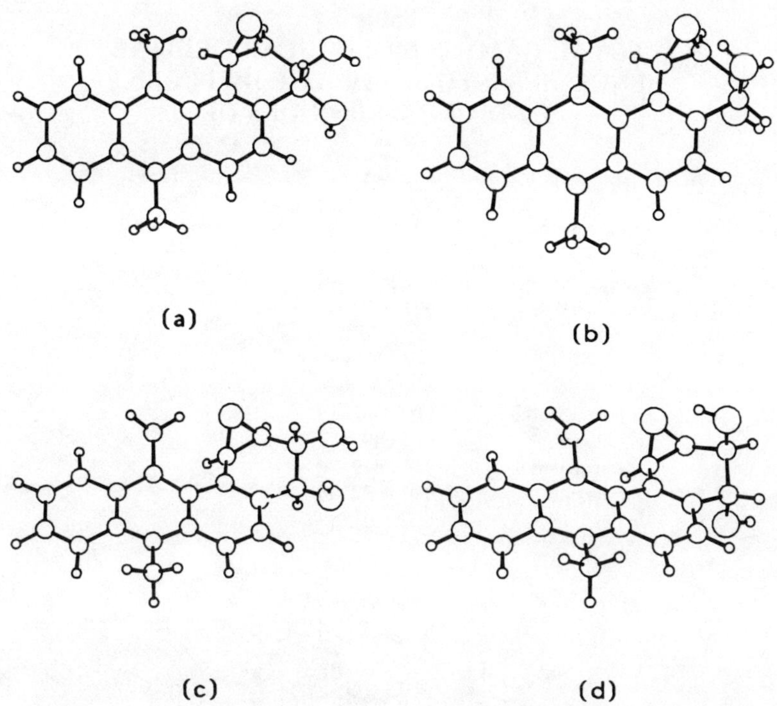

FIGURE 12. (a) Anti-diequatorial diol-epoxide of DMBA; (b) syn-diaxial diol-epoxide of DMBA; (c) syn-diequatorial diol-epoxide of DMBA; and (d) anti-diaxial diol-epoxide of DMBA.

Figure 11 shows the fragment used to determine the hydroxyl group orientation for the syn-diaxial diastereomer of the bay-region diol-epoxide of DMBA. In the figure we show the final O–H orientations of the diol. It is of interest that the benzylic hydroxyl group, the one opposite the epoxide points towards the epoxide. Apparently a hydrogen bond is formed between the hydrogen atom of this group and the epoxide. Such hydrogen bond formation had originally been postulated to enhance the reactivity of the syn-diastereomer with respect to the anti-diastereomer of the bay-region diol epoxides of BP.[62] More recently, such conformation has been demonstrated by X-ray crystallographic studies of 3,4-epoxy-2-methyl-1,2,3,4-tetrahydro-1-naphthol.[63]

The hydroxyl group orientations found for each of the fragments investigated were then used in constructing the final total molecular geometries of the different diol-epoxides of DMBA. These complete molecular structures are shown in Figure 12. These total structures were then used as input to the Gamess molecular orbital program to determine the total energy of each conformer of the DMBA diol-epoxide diastereomers.

Table 3 lists the total Gamess molecular orbital energies for the bay-region diol-epoxides of BA and DMBA. One notes for BA, that the AE conformer is calculated to be somewhat more stable than the AD conformer (by approximately 0.1 eV.), whereas the SD and SE conformers are calculated to be approximately equal in energy. Comparing these results with the results for DMBA one notes that the relative stabilities of the different conformers of the anti- and syn-diastereomers shift towards increased stability of the uncommon type of conformer as one goes from BA to DMBA. Namely, the SE and AD conformers become more stable with respect to the SD and AE conformers. Also, these shifts are approximately comparable and presumably due to destabilization of the SD and AE conformers due to steric crowding in the bay-region.

If one expresses the energies of the various conformers as follows:

Table 3
TOTAL GAMESS MOLECULAR ORBITAL ENERGIES (STO-3G) OF THE DIFFERENT 1,2-DIOL-3,4-EPOXIDES OF

BA	
Anti-diequatorial	−902.9243 Au. (atomic energy units)
Syn-diaxial	−902.9226 Au.
Syn-diequatorial	−902.9222 Au.
Anti-diaxial	−902.9199 Au.
DMBA	
Anti-diequatorial	−980.0501 Au.
Syn-diaxial	−980.0492 Au.
Syn-diequatorial	−980.0519 Au.
Anti-diaxial	−980.0492 Au.

1 Au. = 27.2 eV.

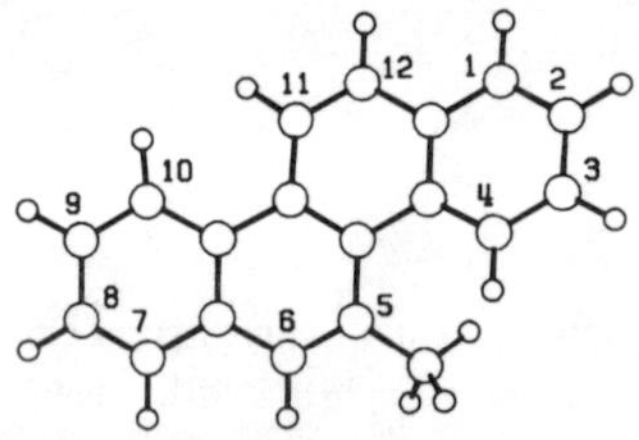

FIGURE 13. 5-Methylchrysene (5-MeC).

$$E(AE) = E_o - \delta R + \delta S - \delta H$$

$$E(SD) = E_o - \delta R + \delta S$$

$$E(SE) = E_o - \delta H$$

$$E(AD) = E_o$$

δR is the increment of binding energy due to the favorable ring conformation of the AE and SD conformers. δS is the increment of energy due to the steric crowding arising from a methyl group in the bay region, and δH is an increment of energy associated with conformers having a diequatorial diol. Then with the further assumption that the magnitudes of all of the δs are comparable, one obtains a set of energies with relative values consistent with those given in Table 3.

C. 5-Methylchrysene (5-MeC)

5-Methylchrysene (Figure 13) is another PAH that has been of interest and extensively studied, particularly by the research group at the American Health Foundation.[29,64-71] This molecule is of interest since it is one additional example of a PAH that has been activated by methylation in the bay-region. Interest in this PAH is also not purely academic, since it has been found as one of the products in cigarette smoke. 5-MeC is the most tumorigenic of all the chrysene monomethyl isomers.[64,65] It is apparently activated via metabolism to a bay-region diol epoxide.[68,70] The molecule is unique since it is the only tetracyclic PAH

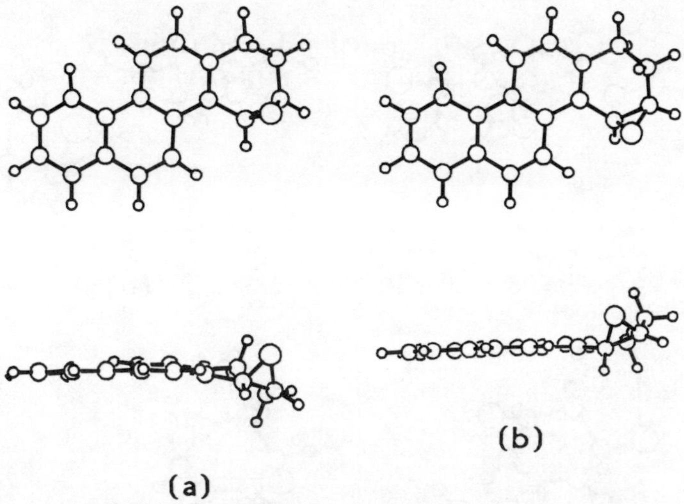

FIGURE 14. Epoxide conformers of chrysene. (a) Perp and (b) nonperp.

that contains two bay-regions. Methylation can therefore be in either of the two inequivalent bay-regions of a bay-region diol-epoxide, i.e., the methyl group can be in the same bay-region that the epoxide is in, or it can be in the other bay-region. Recent work[71,72] has focused on these two isomeric diol-epoxides in an attempt to determine which of these is the ''ultimate carcinogen'' of 5-MeC.

We have calculated the total molecular orbital energies for all of the monomethyl isomers of the perp (Figure 14a) and nonperp (Figure 14b) epoxides of chrysene. Similar to the the other epoxides we have treated, the perp bay-region epoxide conformer of chrysene has a benzo-ring geometry that is similar to the ring geometry of the usually observed diastereomers, namely, the syn-diaxial (Figure 15b) and anti-diequatorial (Figure 15a) bay-region diol epoxides of chrysene. The nonperp bay-region epoxide conformer has a benzo-ring geometry that is similar to the ring geometry of the two conformers that are not generally observed, namely, the syn-diequatorial (Figure 15c) and anti-diaxial (Figure 15d) bay-region diol-epoxides of chrysene.

Table 4 gives the total molecular orbital energies of the two different bay-region epoxide conformers, perp and nonperp, for each of the isomeric methylated isomers that do not involve methylation on the active benzo-ring. The molecular orbital energies are also given for the unsubstituted conformers. As described in Section II, input geometries to the Gaussian 70 *ab initio* molecular orbital program[57] and INDO molecular orbital program[54] have been determined with use of the Allinger force-field program.[50]

One notes from examination of the INDO energies of Table 4, that except for the isomer with the methyl group in the same bay-region as the epoxide, all of the perp conformers are lower in energy than the nonperp conformers. The Gaussian 70 energies exhibit a similar trend although the energy differences are smaller and the ordering is not absolute. These results suggest that the effect of methylation in the bay-region, adjacent to the epoxide, will destabilize both syn-diaxial and anti-diequatorial diastereomers with respect to the syn-diequatorial and anti-diaxial conformers. This result is similar to what we have previously found and discussed in connection with DMBA.

Recent measurements[71] do show that the ''catalytic effect'' of DNA in enhancing the rates of hydrolysis of dihydrodiol epoxides of 5-MeC is consistent with such destabilization for the syn-isomer. The presence of DNA enhances the rate of hydrolysis of the syn-diol-epoxide with both epoxide and methyl group in the same bay-region over that of the isomer

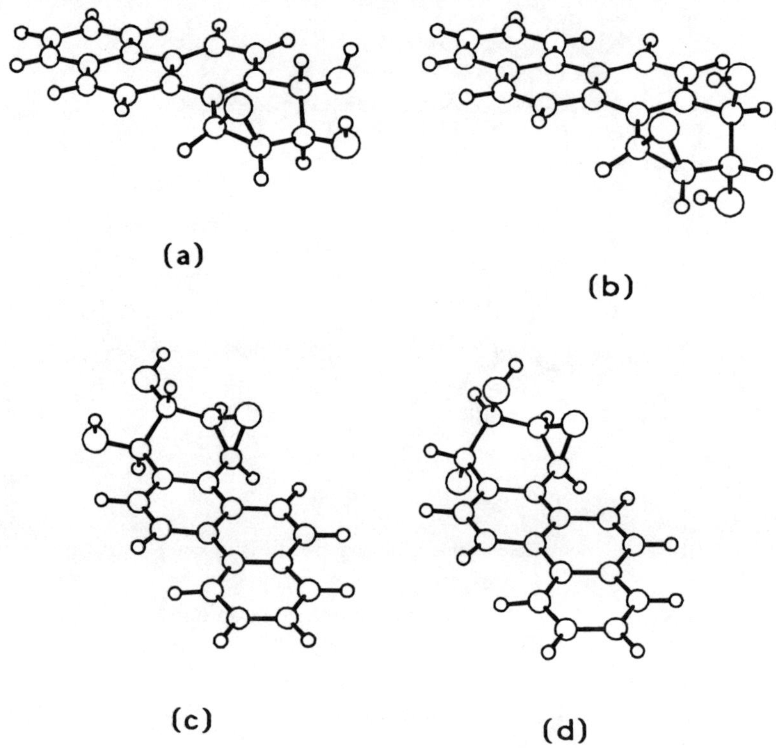

FIGURE 15. (a) Anti-diequatorial diol-epoxide of chrysene; (b) syn-diaxial diol-epoxide of chrysene; (c) syn-diequatorial diol-epoxide of chrysene; and (d) anti-diaxial diol-epoxide of chrysene.

Table 4
MOLECULAR ORBITAL ENERGIES OF THE BAY-REGION EPOXIDES OF CHRYSENE AND METHYLATED CHRYSENES

	INDO	Gaussian-70 (STO-3G)
Perp	−151.87925 Au. (0.065 eV)	−755.28913 Au.
Nonperp	−151.87685	−755.28948(0.0095 eV)
6-Perp	−160.32106(0.079)	−793.86879
6-Nonperp	−160.31816	−793.86940(0.017)
7-Perp	−160.32273(0.12)	−793.86994(0.022)
7-Nonperp	−160.31818	−793.86913
8-Perp	−160.32353(0.10)	−793.87293
8-Nonperp	−160.31977	−793.87306(0.0035)
9-Perp	−160.32246(0.090)	−793.87203
9-Nonperp	−160.31925	−793.87281(0.021)
5-Perp	−160.30854	−793.85405
5-Nonperp	−160.31009(0.042)	−793.85554(0.041)
10-Perp	−160.30919(0.10)	−793.85423(0.018)
10-Nonperp	−160.30546	−793.85357
11-Perp	−160.30848(0.11)	−793.85440(0.049)
11-Nonperp	−160.30437	−793.85260

Note: Total molecular orbital energies are given in atomic units (Au.). Energy differences between the two conformers are given in electronvolts and appear in parenthesis. This value is placed next to the more stable of the two conformers.

Table 5
DELOCALIZATION ENERGY OF THE BAY-REGION (3,4) EPOXIDE OF CHRYSENE AND METHYLATED CHRYSENE

	INDO	Gaussian 70 (STO-3G)
Chrysene	0.384 Au.	0.411 Au.
11-MeC	0.397	0.418
5-MeC	0.392	0.416
9-MeC	0.385	0.414
10-MeC	0.388	0.413
8-MeC	0.384	0.412
7-MeC	0.384	0.412
6-MeC	0.385	0.412

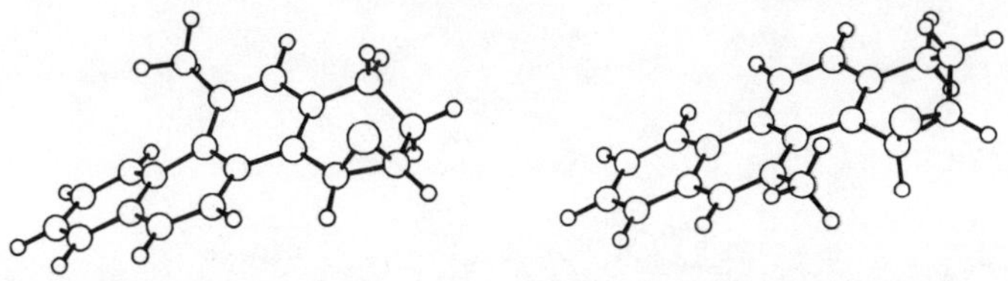

(a) (b)

FIGURE 16. (a) 3,4-Epoxide of 11-MeC and (b) 3,4-epoxide of 5-MeC.

with the epoxide and methyl group in separate bay-regions. One might argue that this might be expected if the presence of the methyl group adjacent to the epoxide were to force a diequatorial diol conformation. Such conformation might be expected to enable the molecule to intercalate more easily, initially, prior to covalently interacting with DNA. Application of such argument to the anti-isomer, however, would not be consistent with the experimental results found,[71,72] since the catalytic effect of DNA, apparently, also enhances the hydrolysis rate of this isomer with both methyl group and epoxide in the same bay-region. Steric crowding in the bay-region of the anti-isomer might therefore be insufficient to yield a diaxial conformation of the diol.

We have also calculated the carbocation delocalization energies associated with methylation at each of the positions that are not on the active benzo-ring. These results are given in Table 5. As previously described, the delocalization energy is the difference in total molecular energy between the initial diol-epoxide and ring-opened carbocation. This should provide a measure of relative ease of epoxide ring opening, or, in other words, a measure of relative epoxide reactivity. Actually, one would like to calculate the energy required for the epoxide to surmount the energy barrier on its way to forming the carbocation. This would necessitate calculating the ''transition state energy''. Since the chemical reaction pathway to the carbocation is not clearly understood, this could only be done with difficulty. Instead, one assumes that the energy differences between the final and initial states will parallel the energy differences between the transition state and initial state. This is essentially, the Hammond postulate[73] used widely in predicting the relative ease of different chemical reaction pathways.

One notes from examination of Table 5 that of the methylated epoxide chrysene isomers, 11-MeC is expected to be the most reactive. This is the isomer with the methyl group and the epoxide in separate bay-regions (Figure 16a). The second most reactive isomer is the

Table 6
MAGNITUDE OF THE AMPLITUDE OF THE LOWEST UNOCCUPIED MOLECULAR ORBITAL (LUMO) OF THE BAY-REGION CARBOCATION OF CHRYSENE

Position (see Figure 17)	Magnitude of the INDO LUMO
5	0.00
6	0.26
7	0.11
8	0.00
9	0.12
10	0.00
11	0.44
12	0.00

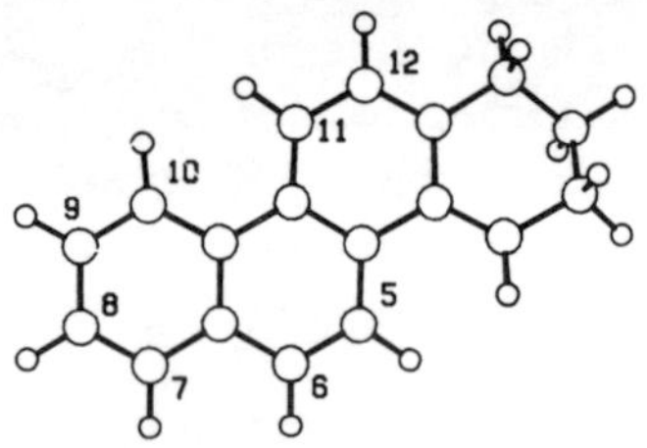

FIGURE 17. Bay-region carbocation of chrysene.

one with the methyl group and epoxide in the same bay-region (Figure 16b). One might have attempted to predict these results from Simple Huckel considerations.[38,39] Table 6 gives the INDO magnitudes of the amplitude of the lowest unoccupied molecular orbital (LUMO) of the unsubstituted chrysene carbocation. This carbocation is shown in Figure 17. For simplicity we have saturated the bonds on the benzo-ring with only hydrogen atoms and neglected the hydroxyl groups. One can argue, that since the effect of a methyl group is to ''release electrons'' that stabilize the carbocation, this will be done most effectively when substitution is at the position of greatest LUMO amplitude. The methyl group can then dump its electrons most effectively into this empty orbital. Examination of Table 6 does indeed show the origin of the high reactivity calculated for the isomer with the methyl group and epoxide in two separate bay-regions. Note that position 11 has the largest LUMO amplitude associated with it. What is surprising, however, is that the isomer calculated to be the second most reactive is the isomer with the methyl group and epoxide in the same bay-region. Examination of Table 6 shows that there is zero amplitude at position 5 and hence we would expect little if any carbocation stabilization at all. Indeed, one can show that a Simple Huckel calculation would yield such result. One, therefore, surmises that the reason for the enhanced reactivity or delocalization energy calculated for 5-MeC has to do with direct interactions between the methyl group and carbocation center. Such interactions are neglected by the Simple Huckel calculation. It is of interest that measurements of the solvolytic reactivity of the two different isomers with both methyl groups and epoxide in the same and different bay-regions show that the latter is more reactive.[71,72] This is consistent with the results just presented. The former isomer is, however, found to be more tumorigenic and to bind more strongly to DNA than the latter isomer.[71,72]

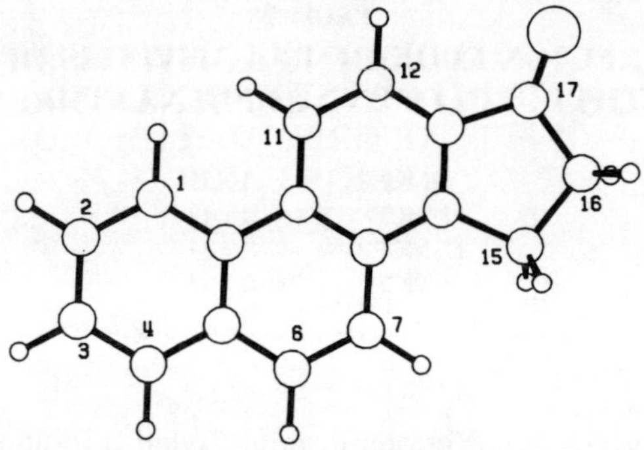

FIGURE 18. 15,16-Dihydrocyclopenta(*a*)phenanthrene-17-one (CPP).

We have also performed calculations for 12-methylbenzathracene (12-MBA) and 11-methylcyclopentaphenanthrene (11-MCPP).[55] The latter will be discussed in the next section. A simple generalization that one might make after examining the calculated results for 5-MeC, 12-MBA, and 11-MCPP, is that, whereas, the isomer methylated in the bay-region is expected to be a reactive isomer on the basis of the delocalization calculations, the isomer obtained by methylating the PAH at the position where the LUMO of the bay-region carbocation is the largest should exhibit the greatest delocalization energy. One therefore expects this isomer to exhibit the greatest solvolytic reactivity of all the isomers.

D. 15,16-Dihydro-11-methylcyclopenta(*a*)phenanthren-17-one (MCPP)

CPP (Figure 18) and its methylated isomers are one of a number of cyclopenta(*a*)phenanthrenes that have been studied for their ability to induce carcinogenesis.[74] This series has been of interest since it is structurally related to both polycyclic aromatic hydrocarbons and steroids. The system has been studied extensively over the years by Coombs and co-workers as well as up to the present.[75,76]

Of the systems studied with different substituents on the five-membered ring and methylated at different positions on the six-membered rings, the most carcinogenic molecule has been found to be 15,16-dihydro-11-methylcyclopenta(*a*)phenanthrene-17-one (11-MCPP). This is the methylated isomer with the methyl group at position 11, in the bay-region of the molecule, shown in Figure 18. Hence, this is yet another example of the activation of PAH by bay-region methylation. It is of interest that the results of the studies on 11-MCPP and its metabolites parallel what has been found for molecules such as 5-MeC and DMBA. Namely, it is the bay-region diol-epoxide, i.e., the 3,4-diol-1,2-epoxide of 11-MCPP which is apparently the "ultimate carcinogen" or that particular reactive species predominantly responsible for binding to DNA and for the initiation of carcinogenesis. In vitro metabolism studies have shown that except for methylation on the active benzo- or "A-ring", other methylated species yield similar amounts of the 3,4-diol.[75] One might presume this diol to be the proximate carcinogen of other methylated species as well as of 11-MCPP. While this remains to be established, it is a reasonable working hypothesis to assume that the bay-region diol-epoxide is the "ultimate carcinogen" of each of the methylated isomers of CPP.

With this in mind, Table 7 shows the calculated bay-region epoxide reactivity or bay-region carbocation delocalization energy for each of the methylated isomers of CPP. The effect of methylation is given for only four substitutional sites. We have not treated methylation on the active "A-ring" nor on the five-membered-ring. It is of interest that methylation at position 7 yields the greatest carbocation delocalization energy. This is consistent with the large LUMO amplitude calculated at this position for the carbocation of CPP. The

Table 7
BAY REGION EPOXIDE REACTIVITIES OF THE MONOMETHYLCYCLOPENTA(*a*)PHENANTHREN-17-ONES

7-MCPP	10.55[a]
11-MCPP	10.42
12-MCPP	10.35
6-MCPP	10.23

[a] Electronvolts.

LUMO amplitude at this position is greater than that found at positions 11 or 12. The isomer with methyl group at the peri position, position 6, has not been treated for reasons stated previously, namely the expected inhibition of important metabolic reactive products by the presence of the methyl group. One expects little conversion of the parent to a bay-region diol-epoxide for the peri-substituted isomer.[45] It is of further interest that the delocalization energy for 11-MCPP is calculated to be greater than that calculated for 12-MCPP. Again, as previously argued, position 11, the position in the bay-region would have zero amplitude in the Huckel approximation. Position 12 should have nonvanishing amplitude. Therefore, there is a through space interaction that apparently stabilizes the bay-region carbocation of 11-MCPP. All of this is completely analogous to what we have found for 5-MeC and 12-methylbenzanthracene.[55]

It is of interest that 11-MCPP is more carcinogenic than 7-MCPP.[74,75] In the same manner that one finds the 1,2-diol-3,4-epoxide of 5-MeC more carcinogenic and more reactive with DNA than the 1,2-diol-3,4-epoxide of 11-MeC, but the solvolytic reactivities reversed;[72] it would be of interest to see if the bay-region diol-epoxide of 7-MCPP was more solvolytically reactive than the bay-region diol-epoxide of 11-MCPP. Perhaps, the greatest solvolytic reactivity of the methylated isomers of bay-region diol-epoxides is achieved by methylation at the position where the carbocation LUMO amplitude is the greatest. Certainly this is what one would expect. On the other hand, LUMO amplitude is certainly an incorrect and misleading index for ranking the induction of carcinogenicity by the bay-region methylated isomer with respect to the other isomers.

Finally, if one compares the total molecular orbital energies of the perp and nonperp 1,2-epoxide conformers with and without the methyl group in the bay-region, at position 11, one finds a similar result to what we have found in connection with DMBA and 5-MeC; namely the perp conformer is destabilized with respect to the nonperp conformer upon methylation in the bay-region. As previously discussed, one might infer from this result that methylation in the bay-region of CPP should destabilize both anti-diequatorial and syn-diaxial diastereomers with respect to the anti-diaxial and syn-diequatorial conformers.

IV. CONCLUSIONS

In summary, we have examined certain effects of intramolecular steric crowding on factors known to hold important consequences for the induction of PAH carcinogenesis. We have shown that methylation in the bay-region of PAH, known to significantly activate PAH, should be expected to enhance the ease of diol-epoxide ring opening with respect to that of the parent molecule. Such ring opening is presumably expected to be a crucial rate-limiting step in the induction of PAH carcinogenesis. Furthermore, it has been shown that bay-region methylation of PAH should be expected to induce changes in the relative conformational stability of diol-epoxide diastereomers believed to be intimately involved in the induction of PAH carcinogenesis.

On a more general note, the present work has attempted to illustrate one way in which the computer can be used to enhance our understanding of complicated biological events. For such processes, dependent upon numerous parameters, many of which are incompletely understood, it is necessary that computational simulation go hand in hand with the detailed results of experiments. The results of such experiments have clearly shown why a single number or set of numbers will not provide an absolute ranking of the carcinogenic potency of all PAH.[77] Taken together with the results of experiment, however, the numbers, should enhance our understanding of important biological processes and hopefully provide us with some interesting hypotheses to check out with further experiments.

REFERENCES

1. **Arcos, J. C. and Argus, M. F.,** *Chemical Induction of Cancer,* Academic Press, New York, 1974, 5.1.1.6.
2. **Pullman, B.,** Recent developments on the mechanism of chemical carcinogenesis by aromatic hydrocarbons, *Int. J. Quantum Chem.,* 16, 669, 1979.
3. **Lowe, J. P. and Silverman, B. D.,** Predicting carcinogenicity of polycyclic aromatic hydrocarbons, *Acc. Chem. Res.,* 17, 332, 1984.
4. **Buening, M. K., Levin, W., Wood, A. W., Chang, R. L., Lehr, R. E., Taylor, C. W., Yagi, H., Jerina, D. M., and Conney, A. H.,** Tumorigenic activity of benzo(e)pyrene derivatives on mouse skin and in newborn mice, *Cancer Res.,* 40, 203, 1980.
5. **Lowe, J. P.,** Theoretical study of relations between bay-region and K-region indices of carcinogenicity, *Int. J. Quantum Chem. Symp.,* 9, 5, 1982.
6. **Lowe, J. P.,** Theoretical analysis of K-region and bay-region indicators of carcinogenicity, *Int. J. Quantum Chem.,* 22, 1221, 1982.
7. **Brookes, P. and Lawley, P. D.,** Evidence for the binding of polynuclear aromatic hydrocarbons to the nucleic acids of mouse skin. Relationship between carcinogenic power of hydrocarbons and their binding to deoxyribonucleic acid, *Nature (London),* 202, 781, 1964.
8. **Ashurst, S. W., Cohen, G. M., Newnow, S., DiGiovanni, J., and Slaga, T. J.,** Formation of benzo(a)pyrene/DNA adducts and their relationship to tumor initiation in mouse epidermis, *Cancer Res.,* 43, 1024, 1983.
9. **Borgen, A., Darvey, H., Castagnoli, N., Crocker, T. T., Rasmussen, R. E., and Yang, I. Y.,** Metabolic conversion of Benzo(a)pyrene by Syrian hamster liver microsomes and binding of metabolites to deoxyribonucleic acid, *J. Med. Chem.,* 16, 502, 1973.
10. **Sims, P., Grover, P. L., Swaisland, A., Pal, K., and Hewer, A.,** Metabolic activation of benzo(a)pyrene proceeds by a diol-epoxide, *Nature (London),* 252, 326, 1974.
11. **Phillips, H. P. and Sims, P.,** Polycyclic aromatic hydrocarbon metabolites: their reactions with nucleic acids (and other articles), in *Chemical Carcinogens and DNA,* Vols. 1 and 2, Grover, P. L., Ed., CRC Press, Boca Raton, Fla., 1979.
12. **Conney, A. H.,** Induction of microsomal enzymes by foreign chemicals and carcinogenesis by polycyclic aromatic hydrocarbons: G. H. A. Clowes Memorial Lecture, *Cancer Res.,* 42, 4875, 1982.
13. **Jerina, D. M., Lehr, R. E., Yagi, H., Hernandez, O., Dansette, P. M., Wislocki, P. G., Wood, A. W., Chang, R. L., Levin, W., and Conney, A. H.,** Mutagenicity of benzo(a)pyrene derivatives and the description of a quantum mechanical model which predicts the ease of carbonium ion formation from diol-epoxides, in *In Vitro Metabolic Activation and Mutagenesis Testing,* de Serres, F., et al., Eds., Elsevier, Amsterdam, 1976, 159.
14. **Jerina, D. M., Sayer, J. M., Thakker, D. R., Yagi, H., Levin, W., Wood, A. W., and Conney, A. H.,** Carcinogenicity of polycyclic aromatic hydrocarbons: the bay-region theory, in *Carcinogenesis: Fundamental Mechanisms and Environmental Effects,* Pullman, B., Ts'o, P. O. P., and Gelboin, H., Eds., D. Reidel Publishing, Dortrecht, 1980, 1.
15. **Weinstein, I. B., Jeffrey, A. M., Jennette, K. W., Blobstein, S. H., Harvey, R. G., Harris, C., Autrup, H., Kasai, H., and Nakanishi, K.,** Benzo(a)pyrene diol epoxides as intermediates in nucleic acid binding in vitro and in vivo, *Science,* 193, 592, 1976.
16. **Weinstein, I. B., Yamasaki, H., Wigler, M., Lee, L., Fisher, P. B., Jeffrey, A. M., and Grunberger, D.,** Molecular and cellular events associated with the action of initiating carcinogens and tumor promoters, in *Carcinogens: Identification and Mechanisms of Action,* Griffin, A. C. and Shaw, C. B., Eds., Raven Press, New York, 1979.
17. **Jerina, D. M. and Daly, J. W.,** Oxidation at carbon, in *Drug Metabolism,* Parke, D. V. and Smith, R. L., Eds., Taylor and Francis, London, 1977, 13.

18. **Jerina, D. M., Yagi, H., Levin, W., and Conney, A. H.,** Carcinogenicity of benzo(a)pyrene, in *Proc. Alfred Benzon Symp., No. 10,* Academic Press, New York, 1977, 261.
19. **Pruess-Schwartz, D., Sebti, S. M., Bruner, G., Gilham, P. T., and Baird, W. M.,** Separation of benzo(a)pyrene-DNA adducts in cells in culture by chromotography on cellulose containing immobilized dihydroxyboryl groups, in *Polynuclear Aromatic Hydrocarbons: Mechanisms, Methods, and Metabolism,* Proc. 8th Int. Symp., Cooke, M. and Dennis, A. J., Eds., Battelle Press, Columbus, Ohio, 1985, 1029.
20. **Geacintov, N. E.,** Reaction pathways of benzo(a)pyrene diol epoxide and DNA. Conformations of adducts, in *Polynuclear Aromatic Hydrocarbons: Physical and Biological Chemistry,* Proc. 6th Int. Symp., Cooke, M., Dennis, A. J., and Fisher, G. L., Eds., Battelle Press, Columbus, Ohio, 1982, 311.
21. **Silverman, B. D.,** Determination of the molecular structure of the bay-region epoxides of benzo(a)pyrene by a force-field calculation, *Cancer Biochem. Biophys.,* 6, 131, 1983.
22. **Neidle, S. and Cutbush, S. D.,** X-ray crystallographic analysis of the (±)-7β,8α-dihydroxy-9β,10β-epoxy-7,8,9,10-tetrahydrobenzo(a)pyrene: molecular structure of a 'syn' diol epoxide, *Carcinogenesis,* 4, 415, 1983.
23. **Sayer, J. M., Yagi, H., Croisy-Delcey, M., and Jerina, D. M.,** Novel bay-region diol epoxides from benzo(c)phenanthrene, *J. Am. Chem. Soc.,* 103, 4970, 1981.
24. **Levin, W., Wood, A. W., Chang, R. L., Ittah, Y., Croisy-Delcey, M., Yagi, H., Jerina, D. M., and Conney, A. H.,** Exceptionally high tumor-initiating activity of benzo(c)phenanthrene bay-region diol-epoxides on mouse skin, *Cancer Res.,* 40, 3910, 1980.
25. **Wood, A. W., Chang, R. L., Levin, W., Ryan, D. E., Thomas, P. E., Croisy-Delcey, M., Ittah, I., Yagi, H., Jerina, D. M., and Conney, A. H.,** Mutagenicity of the dihydrodiols and bay-region diol-epoxides of benzo(c)phenanthrene in bacterial and mammalian cells, *Cancer Res.,* 40, 2876, 1980.
26. **Jerina, D. M., Sayer, J. M., Yagi, H., Croisy-Delcey, M., Ittah, Y., and Thakker, D. R.,** Highly tumorigenic bay-region diol-epoxides from the weak carcinogen benzo(c)phenanthrene, *Adv. Exp. Med. Biol.,* 136(Part A), 501, 1981.
27. **Silverman, B. D. and La Placa, S. J.,** Bay region diol epoxides of benzo(c)phenanthrene: force-field molecular structures, *J. Chem. Soc. Perkin Trans. 2,* 415, 1982.
28. **Silverman, B. D.,** Steric crowding and conformer stability of the diol-epoxides of benzo(c)phenanthrene, *Cancer Biochem. Biophys.,* 6, 23, 1982.
29. **Hecht, S. S., Amin, S., Rivenson, A., and Hoffman, D.,** Tumor initiating activity of 5,11-dimethylchrysene and the structural requirement favoring carcinogenicity of methylated polynuclear aromatic hydrocarbons, *Cancer Lett.,* 8, 65, 1979.
30. **Coombs, M. M., Bhatt, T. S., and Croft, C. J.,** Correlation between carcinogenicity and chemical structure in cyclopenta(a)phenanthrenes, *Cancer Res.,* 33, 832, 1973.
31. **DiGiovanni, J., Diamond, L., Harvey, R. G., and Slaga, T. J.,** Enhancement of the skin tumor-initiating activity of polycyclic aromatic hydrocarbons by methyl-substitution at non-benzo 'bay-region' positions, *Carcinogenesis,* 4, 403, 1983.
32. **Wislocki, P. G., Fiorentini, K. M., Fu, P. P., Yang, S. K., and Lu, A. Y. H.,** Tumor-initiating ability of the twelve monomethylbenz(a)anthracenes, *Carcinogenesis,* 3, 215, 1982.
33. **Sawicki, J. T., Moschel, R. C., and Dipple, A.,** Involvement of both syn- and anti-dihydro-diol-epoxides in the binding of 7,12-dimethylbenz(a)anthracene to DNA in mouse embryo cell cultures, *Cancer Res.,* 43, 3212, 1983.
34. **Meehan, T. and Straub, K.,** Double stranded DNA stereoselectively binds benzo(a)pyrene diol epoxides, *Nature (London),* 277, 410, 1979.
35. **Macleod, M. C. and Zachary, K.,** Catalysis of carcinogen-detoxification by DNA: comparison of enantiomeric diol epoxides, *Chem. Biol. Interact.,* 54, 45, 1985.
36. **Subbiah, A., Islam, S. A., and Neidle, S.,** Molecular modelling studies on the non-covalent intercalative interactions between DNA and the enantiomers of anti-benzo(a)pyrene 7,8-diol-9,10-epoxide, *Carcinogenesis,* 4, 211, 1983.
37. **Miller, K. J., Burbaum, J. J., and Dommen, J.,** Binding of benzo(a)pyrene-diol epoxides and -triol carbonium ions to DNA, in *Polynuclear Aromatic Hydrocarbons: Physical and Biological Chemistry,* Proc. 6th Int. Symp., Cooke, M., Dennis, A. J., and Fisher, G. L., Eds., Battelle Press, Columbus, Ohio, 1982, 515.
38. **Silverman, B. D. and Lowe, J. P.,** Carcinogenicity of methylated hydrocarbons: effect of methylation on the calculated diol-epoxide reactivity, *Cancer Biochem. Biophys.,* 6, 89, 1982.
39. **Silverman, B. D. and Lowe, J. P.,** Diol-epoxide rectivity of methylated polycyclic aromatic hydrocarbons(PAH): Ranking the reactivity of the positional monomethyl isomers, in *Polynuclear Aromatic Hydrocarbons: Physical and Biological Chemistry,* Proc. 6th Int. Symp., Cooke, M., Dennis, A. J., and Fisher, G. L., Eds., Battelle Press, Columbus, Ohio, 1982, 743.
40. **Smith, J. A. and Seybold, P. G.,** Methylbenz(a)anthracenes: correlations between theoretical reactivity indices and carcinogenicity, *Int. J. Quantum Chem. Symp.,* 5, 311, 1981.

41. **Poulsen, M. T. and Loew, G. H.,** Quantum chemical studies of methyl and fluoro analogs of chrysene: metabolic activation and correlation with carcinogenic activity, *Cancer Biochem. Biophys.*, 5, 81, 1981.
42. **Yang, S. K., Chou, M. W., Evans, F. E., and Fu, P. P.,** Metabolism of 8-hydroxymethyl-benz(a)anthracene by rat liver microsomes, stereochemistry of dihydrodiol metabolites and the effect of enzyme induction, *Drug Metab. Disposition,* 12, 403, 1984; and references therein.
43. **Wong, T. K., Chiu, P., Fu, P. P., and Yang, S. K.,** Metabolic study of 7-methylbenzo(a)pyrene with rat liver microsomes: separation by reversed-phase and normal-phase high performance liquid chromatography and characterization of metabolites, *Chem. Biol. Interact.*, 36, 153, 1981.
44. **MacNicoll, A. D., Grover, P. L., and Sims, P.,** The metabolism of a series of polycyclic hydrocarbons by mouse skin maintained in short term organ culture, *Chem. Biol. Interact.*, 29, 169, 1980.
45. **Hamernik, K. L., Chiu, P., Chou, M. W., Fu, P. P., and Yang, S. K.,** Metabolic activation of 6-methylbenzo(a)pyrene, in *Polynuclear Aromatic Hydrocarbons: Formation, Metabolism, and Measurement,* Proc. 7th Int. Conf., Cooke, M. and Dennis, A. J., Eds., Battelle Press, Columbus, Ohio, 1983, 583.
46. **Yang, S. K., Chou, M. W., and Fu, P. P.,** Metabolic and structural requirements for the carcinogenic potencies of unsubstituted and methyl-substituted polycyclic aromatic hydrocarbons, in *Carcinogenesis: Fundamental Mechanisms and Environmental Effects,* Pullman, B., Ts'o, P. O. P., and Gelboin, H., D. Reidel Publ., Dortrecht, 1980, 143.
47. **Yates, K.,** *Huckel Molecular Orbital Theory,* Academic Press, New York, 1978.
48. **Hoffman, R.,** An extended Huckel theory, *J. Chem. Phys.*, 39, 1397, 1963; Quantum Chemistry Program Exchange, Program No. 344, Indiana University.
49. I would like to thank Professor A. R. Rossi, Chemistry Dept. University of Connecticut, Storrs, for making this program available to me.
50. **Allinger, N. L. and Sprague, J. T.,** Calculation of the structure of hydrocarbons containing delocalized electronic systems by the molecular mechanics method, *J. Am. Chem. Soc.*, 95, 3893, 1973; Quantum Chemistry Program Exchange, Program No. 318, Indiana University.
51. **Dougherty, D. A. and Mislow, K.,** A combination of empirical force field and extended Huckel molecular orbital calculations as a computational approach to conformational analysis, *J. Am. Chem. Soc.*, 101, 1401, 1979.
52. **Hehre, W., Stewart, R. F., and Pople, J. A.,** Self consistent molecular-orbital methods I. Use of Gaussian expansions of slater type atomic orbitals, *J. Chem. Phys.*, 51, 2657, 1969.
53. **Dupuis, M., Spangler, D., and Wendelowski,** *QCPE,* 1, 91, 1981.
54. **Pople, J. A. and Beveridge, D. L.,** *Approximate Molecular Orbital Theory,* McGraw Hill, New York, 1979; Quantum Chemistry Program Exchange, Program No. 141, Indiana University.
55. **Silverman, B. D.,** Diol-epoxide conformer stability of polycyclic aromatic hydrocarbons (PAH): the effect of a bay-region methyl group, in *Polynuclear Aromatic Hydrocarbons: Formation, Metabolism, and Measurement,* Proc. 7th Int. Symp., Cooke, M. and Dennis, A. J., Eds., Battelle Press, Columbus, Ohio, 1983, 1113.
56. **Johnson, C. K.,** OR TEP-II: A Fortran Thermal-Ellipsoid Plot Program For Crystal Structure Illustrations, Oak Ridge National Laboratory Rep. ORNL-5138, Oak Ridge, Tenn., 1976.
57. Quantum Chemistry Program Exchange, Program No. 236, Indiana University, Bloomington.
58. **Slaga, T. J., Huberman, E., DiGiovanni, J., and Gleason, G.,** The importance of the "bay-region" diol-epoxide in 7,12-dimethylbenz(a)anthracene skin tumor initiation and mutagenesis, *Cancer Lett.*, 6, 213, 1979.
59. **Iball, J.,** A refinement of the structure of 9:10-dimethyl-1:2-benzanthracene, *Nature (London),* 201, 916, 1964.
60. **Binkley, J. S., Whiteside, R. A., Krishnan, R., Seeger, R., De Frees, J., Schlegel, H. B., Topiol, S., Kahn, L. R., and Pople, J. A.,** *QCPE,* 10, 416, 1981.
61. I'd like to thank Arthur Greenberg for suggesting this procedure.
62. **Hulbert, P. B.,** Carbonium ions as ultimate carcinogen of polycyclic aromatic hydrocarbons, *Nature (London),* 256, 146, 1975.
63. **Glusker, J. P., Zacharias, D. E., Whalen, D. L., Friedman, S., and Pohl, T. M.,** Internal hydrogen bond formation in a syn-hydroxyepoxide, *Science,* 215, 695, 1982.
64. **Hoffman, D., Bondinell, W. E., and Wynder, E. L.,** Carcinogenicity of methylchrysenes, *Science,* 183, 215, 1974.
65. **Hecht, S. S., Bondinell, W. E., and Hoffman, D.,** Chrysene and methylchrysenes: presence in tobacco smoke and carcinogenicity, *J. Natl. Cancer Inst.*, 53, 1121, 1974.
66. **Hecht, S. S., Loy, M., Mazzarese, R., and Hoffman, D.,** Synthesis and mutagenicity of modified chrysenes related to the carcinogen, 5-methylchrysenes, *J. Med. Chem.*, 21, 38, 1978.
67. **Hecht, S. S., LaVoie, E., Mazzarese, R., Amin, S., Bedenko, V., and Hoffman, D.,** 1,2-Dihydro-1,2-dihydroxy-5-methylchrysene, a major activated metabolite of the environmental carcinogen, 5-methylchrysene, *Cancer Res.*, 38, 2191, 1978.

68. **Hecht, S. S., Rivenson, A., and Hoffman, D.,** Tumor initiating activity of dihydrodiols formed metabolically from 5-methylchrysene, *Cancer Res.*, 40, 1396, 1980.
69. **Melikian, A. A., LaVoie, E. J., Hecht, S. S., and Hoffman, D.,** Influence of a bay-region methyl group on formation of 5-methylchrysene dihydrodiol epoxide: DNA adducts in mouse skin, *Cancer Res.*, 42, 1239, 1982.
70. **Melikian, A. A., LaVoie, E. J., Hecht, S. S., and Hoffman, D.,** 5-Methylchrysene metabolism in mouse skin epidermis in vivo, diol-epoxide-DNA adduct persistence, and diol-epoxide reactivity with DNA as potential factors influencing the predominance of 5-methylchrysene-1,2-diol-3,4-epoxide-DNA adducts in mouse epidermis, *Carcinogenesis,* 4, 843, 1983.
71. **Melikian, A. A., Leszczynska, J. M., Amin, S., Hecht, S. S., Hoffman, D., Pataki, J., and Harvey, R. G.,** Rates of hydrolysis and extents of DNA binding of 5-methylchrysene dihydrodiol epoxides, *Cancer Res.,* 45, 1990, 1985.
72. **Kim, M. H., Roche, C. J., Geacintov, N. E., Pope, M., Pataki, J., and Harvey, R. G.,** Comparisons of reactivities and structures of covalent adducts derived from the binding of two isomeric bay-region 5-methylchrysene diol epoxides with DNA, submitted to *Chem. Biol. Interact.*
73. **Hammond, G. S.,** A correlation of reaction rates, *J. Am. Chem. Soc.*, 77, 334, 1955.
74. **Coombs, M. M., Bhatt, T. S., Livingston, D. C., Fisher, S. W., and Abbott, P. J.,** Chemical structure, metabolism, and carcinogenicity in the cyclopenta(a)phenanthrene series, in *Polynuclear Aromatic Hydrocarbons: Chemical Analysis and Biological Fate,* Proc. 5th Int. Conf., Cooke, M. and Dennis, A. J., Eds., Battelle Press, Columbus, Ohio, 1981, 63.
75. **Coombs, M. M., Russell, J. C., Jones, J. R., and Ribeiro, O.,** A comparitive examination of the in vitro metabolism of five cyclopenta(a)phenanthrenes of varying carcinogenic potential, *Carcinogenesis,* 6, 1217, 1985.
76. **Russell, J. C., Bhatt, T. S., Jones, J. R., and Coombs, M. M.,** Comparison of the binding of some carcinogenic and non-carcinogenic cyclopenta(a)phenanthrenes to DNA in vitro and in vivo, *Carcinogenesis,* 6, 1223, 1985.
77. **Berenblum, I.,** The problem of quantitative carcinogenesis, in *The Jerusalem Symposia on Quantum Chemistry and Biochemistry: Physico-Chemical Mechanisms of Carcinogenesis,* Proc. Int. Symp., Vol. 1, Bergmann, E. D. and Pullman, B., Eds., Academic Press, New York, 1969.

Chapter 5

CONFORMATIONS OF DNA MODIFIED BY AROMATIC AMINES: MINIMIZED SEMIEMPIRICAL POTENTIAL ENERGY CALCULATIONS AND MODEL BUILDING

S. Broyde and B. E. Hingerty

TABLE OF CONTENTS

I. AROMATIC AMINES AS CARCINOGENS

The first recorded association of cancer with an aromatic amine was made by Rehn in 1895.[1] Rehn noted a cluster of three men with bladder cancer who were working in the same dye factory. One of the raw materials employed was aniline, and the malignancy became known as aniline cancer. It was subsequently revealed that the actual carcinogens in the dye manufacture were 2-naphthylamine and benzidine (4,4′ diaminobiphenyl).[2,3] 1-Naphthylamine, which was also suspect, has proven to be noncarcinogenic when applied pure in dogs.[4] While aniline itself was not the culprit in the dye industry,[3] current evidence suggests that it is weakly carcinogenic in some animal systems.[5] The aromatic amine 4-aminobiphenyl (ABP) is another identified human bladder carcinogen[6] that has been employed in the chemical industry. Yet another well-known aromatic amine, 2-acetylaminofluorene (AAF), is a synthetic substance originally intended for use as an insecticide. It is tumorigenic to the bladder, liver, and other organs of many animals.[7,8] The carcinogenicity of numerous other aromatic amines has been documented.[8] However, many aromatic amines are noncarcinogens, and the carcinogenic potency of others can vary widely. Even chemically very similar substances may have widely different biological effects. The review by Garner et al.[8] summarizes a wealth of information now at hand on aromatic amine carcinogenicity.

II. CRITICAL STEPS IN THE INITIATION OF CANCER BY AROMATIC AMINES

Our current understanding indicates that a number of events may occur following the entry of an aromatic amine into a cell.[9] The chemical interacts with metabolic enzymes to form one or more highly reactive electrophilic species. These then may be detoxified by reaction with water or other substrates including enzymes. Alternatively, the electrophilic moiety may react with DNA to produce one or more covalently linked adducts. These adducts may be repaired prior to replication, and if the repair is error free, no permanent damage ensues. However, if repair is error-prone, or if repair fails to take place altogether, the lesion may cause the replication machinery to err, or it may cause other genetic effects such as chromosomal rearrangement. Furthermore, the adduct might alter gene expression (increasing or decreasing the amount of normal gene product) even without replication and mutation, or chromosomal rearrangement, if it is situated in a control region. Both mutations and overexpression are known to activate oncogenes so that they can induce cell transformation.[10] The conformation of carcinogen-DNA adduct may determine whether a given DNA-adduct is mutagenic or if it exercises its influence by altering the normal control mechanisms. Cell transformation may then eventually ensue as a result of the somatic mutation and/or the altered gene expression control. A detailed current review on chemical carcinogens, including mechanisms of initiation is given in Reference 9.

III. SCOPE OF THE PRESENT WORK

The conformation of the aromatic amine-DNA adduct is likely to play an important part in determining whether a particular adduct is critical to the initiation of cancer. Many solution studies have been carried out, particularly for AAF and its derivatives, which have produced an understanding of the influence that these substances have on DNA conformation (see below). However, surprisingly, atomic resolution data, unambiguously available only through X-ray crystallography, has not yet been obtainable for any aromatic amine-DNA adduct. The difficulty lies both in the synthesis of rigorously pure adducts and in the yet unquantified art of growing suitable crystals. It is only the use of computers that has permitted the generation of molecular views of these adducts.

We have been studying aromatic amine-DNA adducts using minimized semiempirical potential energy calculations to obtain an energy ranked list of conformations for modified deoxydinucleoside monophosphates. The important low energy forms are then investigated in larger polymers by hand model building, computer modeling, and full-scale energy minimization. Our goal has been to study series of chemically related substances whose biological sequelae may differ in order to try to discern some correlation between adduct conformation and biological effect. Our methods and findings will be described below.

IV. METHODS AND THEIR LIMITATIONS

The goal of the work is to obtain a list of minimum energy conformations at atomic resolution, represented by coordinates of all the atoms, together with the free energy associated with each state. To do so, the exact chemical nature of the carcinogen-DNA adduct must be known. Our studies can yield no information on reactivity, dealing as they do only with the conformation of the adduct after reaction. An initial set of adduct coordinates in some arbitrary conformation is generated from crystal structure data if available, or from standard bond lengths and bond angles where no data exists.

The energy of the molecule (an approximation to the relative enthalpy and crudely assuming equal entropy for all states) is then computed with a semiempirical potential energy function that has been devised by Srinivasan and Olson.[11a] Contributions to the energy from Van der Waals, electrostatic, torsional, deoxyribose strain, anomeric gauche effects, and hydrogen bonding[11b] are included. The choice of parameters for these terms together with details on the methods is documented in Reference 12. Each carcinogen requires a new computation of partial charges for the electrostatic term as well as a choice for the rotation barriers at the carcinogen-DNA linkage site. Although this is an important source of conformational flexibility, experimental information about the magnitude of these barriers in solution is sparse. We choose values from chemically analogous systems where no experimental data are at hand, or computed values if available. Counterion condensation is mimicked indirectly by reduction (in the coulombic term) of the partial charge on the nonlinking phosphate oxygens of the DNA. Solvent is included by means of a distance-dependent dielectric function in the coulombic term; this takes cognizance of the fact that the dielectric constant increases, approaching that of water, as the distance between a given atom pair increases so that solvent can come between them. The interaction of each atom with every other atom is computed, excluding those atom pairs whose separation is not dependent on conformation (such as the atoms in a planar base). All single bonds that permit conformational variability are flexible as are the puckerings of the deoxyribose rings of the DNA. However, all bonds lengths and bond angles except those in the deoxyribose, which are the only ones that are in fact quite conformation dependent, are fixed. The energy is then minimized with a modified version of the steepest descent method by Powell.[13] The calculation is repeated for a large number of starting conformations. The lowest energy conformation that is computed for a given

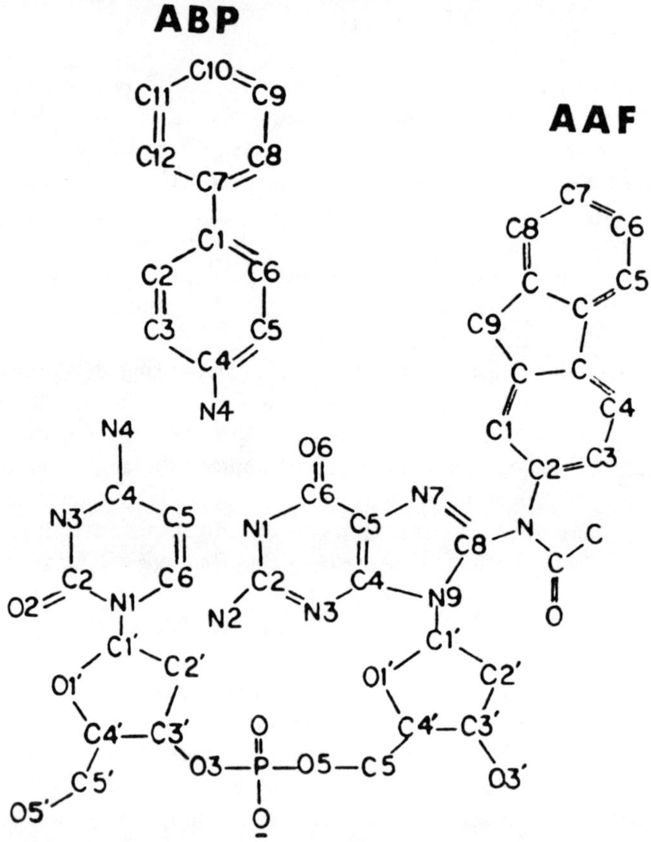

FIGURE 1. Structure, numbering scheme, and conformational angle designations for dCpdG modified at guanine C8 by AAF. In the AF adduct, the $COCH_3$ group is replaced with an H. The guanine N^2 AAF adduct is linked to C3 of the AAF. 4-Aminobiphenyl (inset) links to guanine C8 via its amino group.

substance is termed the global minimum, while other minimum energy forms are called local minima.

Because no reliable method for locating a global minimum or even all of the important local minima has yet been devised, it is appropriate to describe here the search strategy that we employ. The calculations are first carried out on a carcinogen-modified deoxydinucleoside monophosphate (Figure 1). These structures have three advantages. First, they are the conformational building blocks of the A and B forms of DNA, as well as one of the two conformational building blocks of the left-handed Z form.[14,15] Thus, all three polymer forms can be recognized in the dinucleoside monophosphate conformation. Second, modified dimers have only 12 to 15 conformational degrees of freedom. Hence, one can make a fairly thorough, though by no means exhaustive, search of the conformation space of these small DNA subunits on a supercomputer, using optimized code. We are currently making about 4000 trials per adduct, examining all the combinations of rotamer preferences that have been observed for nucleic acids,[14] as well as reasonable combinations for the carcinogen-DNA linkage site. Third, the smaller the number of variables, the better able is the minimization routine to search beyond the nearest local minimum and thus locate novel conformers.

We have been studying the sequence dCpdG. This sequence contains most of the conformational distinctions of the Z helix, permitting discernment of the Z form even on the dimer level, despite the fact that the Z form actually has two different dimeric building blocks.[15] In addition, this sequence is a mutational hot spot and a high reactivity locus for aromatic amines.[16,17]

After obtaining the energy ranked list of dimer conformations, the effect of the important dimer conformers on larger polymer structures is investigated. Three methods are employed for this purpose. (1) We construct hand-built models that incorporate the computed dimer minima into the canonical A, B, or Z helixes, making structural modifications to the standard forms that the model building suggests. While currently less in vogue, it is our view that hand model building often leads to unique structures that would not be so readily apparent with more sophisticated computer technology. It is worth remembering that a model with space filling components is actually a crude energy calculation, including just the repulsive term of the Lennard-Jones potential. (2) Dimer minima may be computationally docked into the standard A, B, and Z type larger polymers, without energy refinement. (3) Dimer minima can be included in larger polymers of any desired conformation, followed by energy minimization of the larger structure. Initially, in this approach, standard helixes are employed as starting conformations for the sites surrounding the carcinogen-modified dimer. During the minimization the entire structure adjusts its conformation as demanded by the lesion. In addition, methods for the generation of novel polymer structures, involving combinations of low energy dimer minima are under investigation.

Our approach differs from other theoretical treatments of DNA polymers which examine larger structures directly. Because of the large number of variables in e.g., a duplex dodecamer, minimizations of such a modified polymer are only capable of locating more or less small distortions to the starting conformation (see, for example, Rao et al.[18] and Pearlman et al.[19]). In our treatment, focus on the search for distortion is made on the dimer level; here the power of the method is adequate to delineate many, though not all, interesting low energy forms. Our method has made successful predictions for unmodified nucleic acid dimers, including the crystal structure of the ribonucleoside monophosphate, GpC,[20] the loop forming conformation[21] subsequently observed in the anticodon and TψC loops of tRNAs,[22] and the low energy DNA backbone conformation[23] subsequently found in the Z form.[15] For carcinogen-modified dimers, we have calculated low energy conformations that yield molecular detail in agreement with solution data where available (see below) and to make predictions where no data exist at all.

Thus, our method, despite inadequacies in the treatment for solvent, salt, entropy, and in the minimization method itself, is capable of making reasonable predictions for nucleic acid dimers, both unmodified and modified by carcinogens, delineating many favored conformations, and to generate feasible larger polymer structures by hand models, computer model building, or energy refinement of the canonical A, B, and Z forms. However, a comprehensive strategy for searching out the conformation space of a duplex dodecamer, for example, with its hundreds of sources of flexibility, to define all completely unknown low energy polymer structures awaits future development.

V. COMPUTED STRUCTURES AND EXPERIMENTAL DATA

A. 2-Acetylaminofluorene (AAF)

1. The Guanine C8 Adduct

a. The Major Adduct of AAF with DNA In Vitro is to Guanine C8 (Figure 1)

This adduct can also be incorporated into *Escherichia coli* by genetic engineering.[16] However, in mammalian systems, the bulk of C8 adducts is enzymatically deacetylated to the 2-aminofluorene derivative (AF).[24,25] Of those AAF adducts remaining, the majority are repaired,[26] while a much smaller fraction persist.[27] Extensive solution studies of the C8-AAF adduct with DNA and its subunits have been carried out for many years;[28-32] these

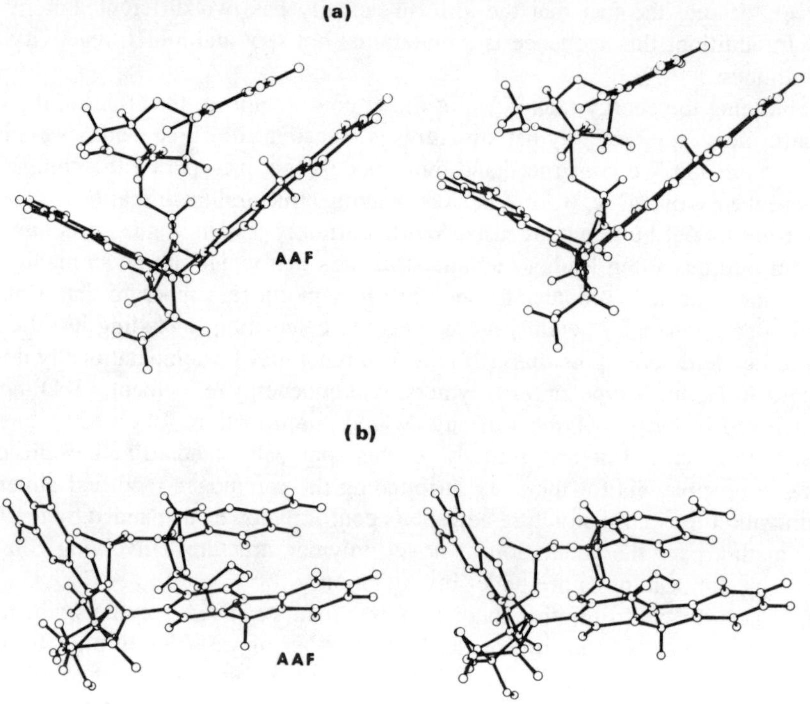

FIGURE 2. Carcinogen-base stacked forms with syn-guanine for the guanine C8 dCpdG-AAF adduct, in stereo view. (a) and (b) are distorted B- and Z-type DNA backbones, respectively.

have led to the understanding that AAF can affect the conformation of DNA at least two ways. In random sequence DNA the data are consistent with a model termed base-displacement[28] or insertion-denaturation,[30] in which the AAF is inserted into the helix by rotation of the guanine from its normal anti conformation in the B helix to syn. This is accompanied by kinking and denaturation. In alternating poly (dG-dC), poly (dG-dC) AAF induces the left-handed Z helix.[29] Because a large amount of solution data are available, albeit without conformational details or a molecular view, the guanine C8-AAF adduct was a reasonable choice to begin our study of aromatic amines, both to test the potentials and to provide an atomic resolution view where none existed.

b. Important Calculated Structures for the dCpdG Adduct[33,38]

1. Carcinogen-Base Stacked Forms with Syn-Guanine

The most important computed conformations, both in terms of energies and number of states, have syn-guanine and stacking of the fluorene ring with adjacent cytidine. A large number of different DNA backbone conformations, including distorted B and Z forms, but not normal B forms, produce this particular morphology. The most favored of these are shown in Figure 2. Some of these conformers are related to one another by the correlated motions, first described by Olson,[34] which the DNA backbone may undergo. The carcinogen is close to perpendicular to the guanine to which it is linked in all the structures, a feature which has been experimentally verified (see below). These forms are in accord with the base displacement or insertion-denaturation model and provide the molecular views that were not previously at hand.

2. Z Forms

Local minimum energy conformations like the dCpdG residue of Z-DNA have also been

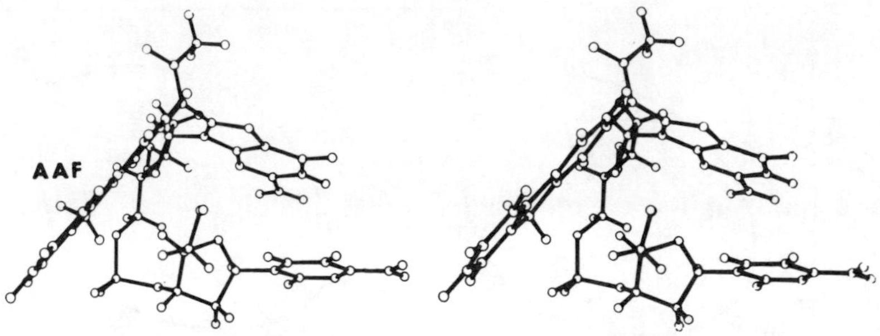

FIGURE 3. Z-DNA type conformation of the guanine C8 dCpdG-AAF adduct, in stereo view. (From Broyde, S. and Hingerty, B., *Carcinogenesis*, 6, 151, 1985. With permission.)

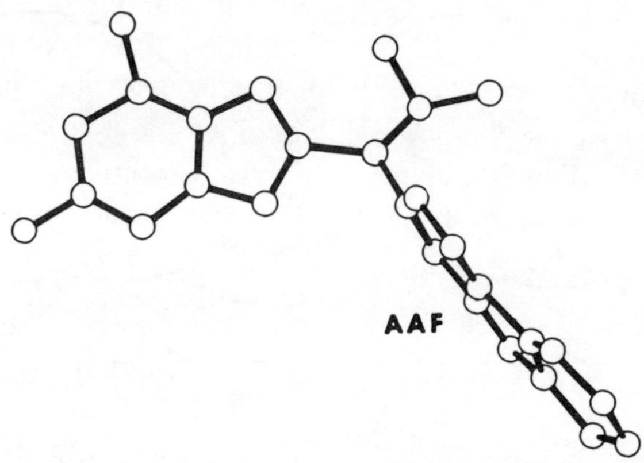

FIGURE 4. Crystal structure of G-AAF.[35]

computed. These undistorted Z forms place the carcinogen at the exterior of the DNA backbone in a flexible orientation (Figure 3). The fluorene ring is again perpendicular to the guanine, which is syn as in Z-DNA.

3. Other Conformations

Base-base stacked forms with syn-guanine and a single base-base stacked structure with anti-guanine (a loop-forming type of conformation) were also computed. It is noteworthy, however, that no unperturbed A or B forms occurred among the computed minima.

c. The Crystal Structure of G-AAF[35]

The only atomic resolution data available to date on AAF linked to a nucleic acid component is the structure of the carcinogen-base adduct solved by Neidle and co-workers.[35] Interestingly, a surprising amount of information was gleanable from this work. The structure itself, shown in Figure 4, reveals the perpendicular orientation between fluorene and guanine that had been predicted in the calculation for the dCpdG adduct. In addition, the carcinogen-base orientation in the crystal proved to be one that is incompatible with carcinogen-base stacking, but that is permitted for Z forms with the carcinogen externally located. This was demonstrated with an energy contour plot in the plane of the carcinogen-base linkages C8-N (α) and N-C2 (β). The map, together with our computed values for α,β in base displaced and in undistorted Z forms, is shown in Figure 5. The results reveal that the Z forms can occupy all the permitted α,β domains, while base displaced forms are much less flexible. The G-AAF crystal structure occurs in a region where only undeformed Z-type conformers are possible.

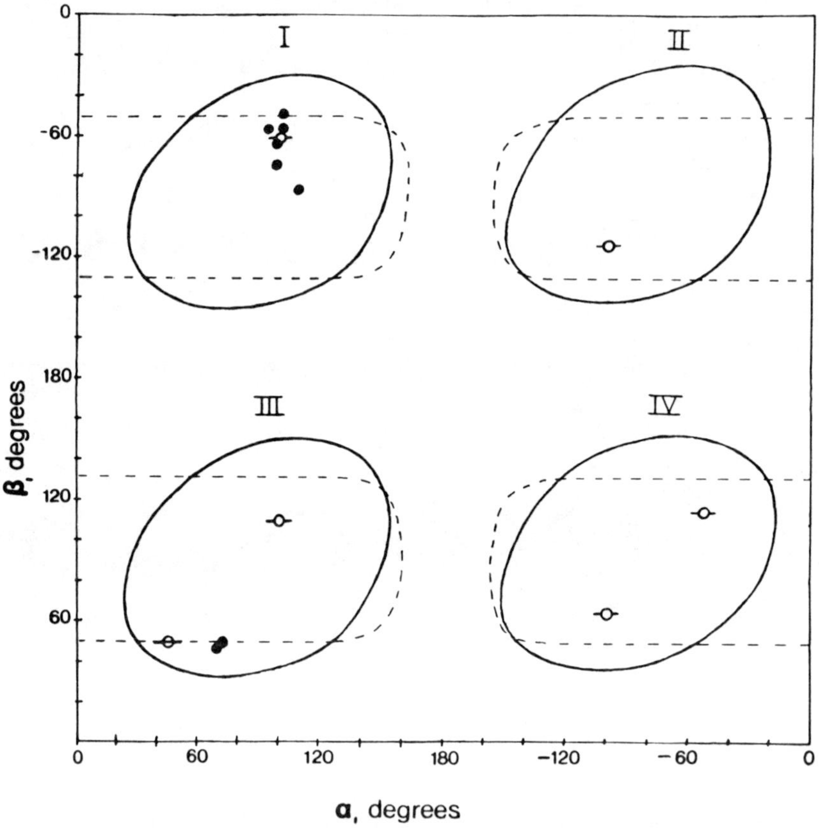

FIGURE 5. Energy contour map in the C8-N(α), N-C2(β) plane for G-AAF. The 5 kcal/mol contour is shown. The solid line is for the cis (0°) domain of the C8-N–C–C torsion, while the broken line is for trans(180°). (●) Designates AAF-base stacked conformers and (-O-) designates Z forms computed for the dCpdG adduct.[33,35,38]

d. Polymer Models

1. AAF-Base Stacking in B Forms

Because base displacement is observed in random sequence DNA which is B form in solution, we modeled two important AAF-base stacked forms into B-DNA. Full energy refinement of the distorted B-like conformer shown in Figure 2a, incorporated into a B form as the starting conformation, yields the structure shown in Figure 6. The AAF is stacked with an opposite base, inducing a bend region where base pairing is not possible, consistent with the experimental observations.

Instead of (or in addition to) adopting a distorted B form, the calculations reveal that AAF-base stacking can be achieved with a distorted Z-form (Figure 2b). Such a local Z-like structure might well be induced by AAF in an overall B helix. A hand model of this conformer, incorporated into a B form modified at the adjacent site to accommodate the distortion, produced a unique structure termed fluorene intrastrand stacking[33] (Figure 7). In this model of the single stranded region induced by the AAF, the wide fluorene ring is stacked with an adjacent cytidine via its innermost six-membered ring, while its distal six-membered ring is intercalated between the next two bases. Again, this model shows the fluorene-base stacking, helix bending, and denaturation observed in solution.

2. Z Forms

i. External Binding

The Z-like dimer conformation shown in Figure 3 can readily be incorporated in a Z helix

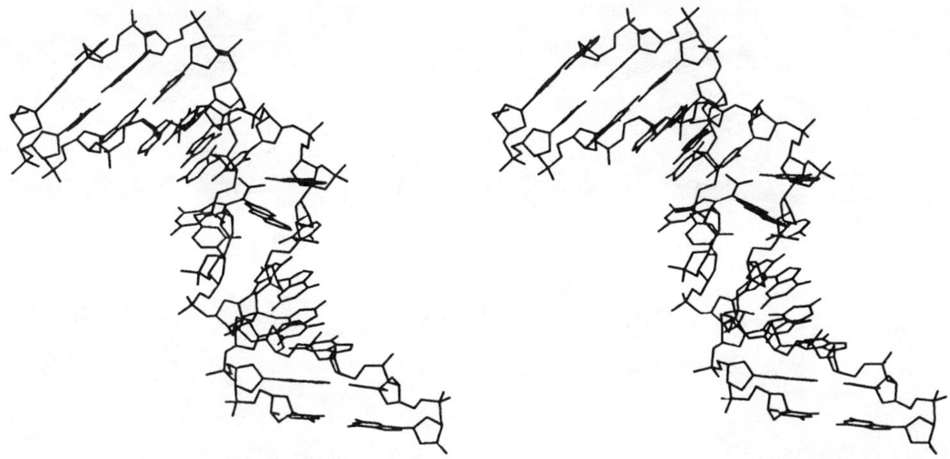

FIGURE 6. Distorted B-type carcinogen-base stacked conformation shown in Figure 2a incorporated into a duplex B form followed by complete energy minimization, in stereo view.

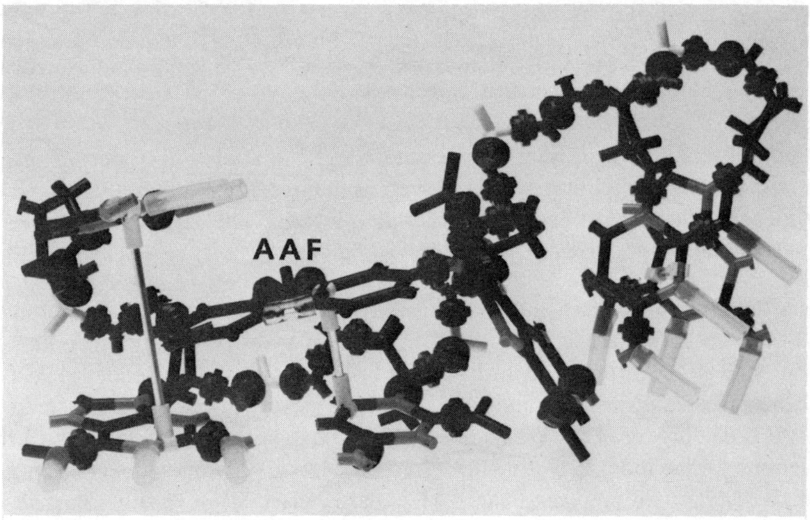

FIGURE 7. Distorted Z-type carcinogen-base stacked conformation shown in Figure 2b built into a modified single stranded B form.[33]

without any distortion. A totally energy-refined Z form incorporating this dimer minimum energy conformation is shown in Figure 8. Model building shows that all basepairs are intact and there are no bends in the helix axis.[33] The AAF resides at the helix exterior, in a flexible position where its orientation can vary. This structure affords one molecular view of the AAF-modified Z-DNA that has been observed in solution.

ii. AAF-Base Stacking in the Z Helix

The distorted Z-like backbone conformation shown in Figure 2b suggests fluorene-base stacking in the Z helix as an additional possibility for AAF-modified Z-DNA. The observation that AAF-modified Z-DNA can recognize two different groups of anti Z-antibodies[36] also indicates the possibility of two distinct conformations. We constructed a model incorporating the distorted Z-like carcinogen-base stacked form of Figure 2b into a Z helix[37] to show that an inserted form is feasible (Figure 9). Some conformational modifications to the adjacent site and the opposite strand are needed to achieve a sterically acceptable model in which

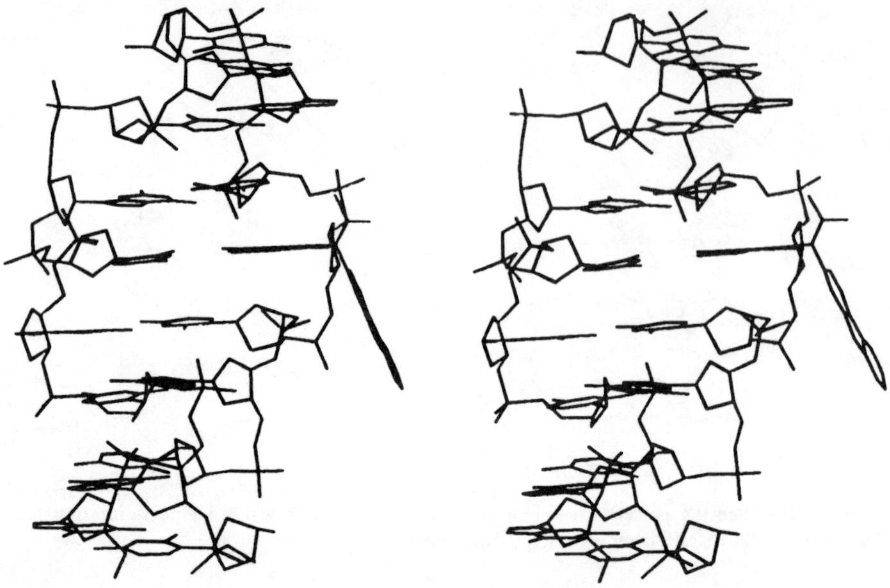

FIGURE 8. Z-type minimum energy conformation shown in Figure 3 incorporated into a Z helix followed by complete energy minimization, in stereo view.

the modified syn-guanine is at the helix exterior and the AAF is inserted. The model is consistent with the antibody binding data as well as the observation that a small but nonzero amount of denaturation occurs in AAF-modified Z-DNA,[29] although other explanations of these data are certainly possible.

3. An Internal Fluorene Model with Syn-Guanine for the 7-Iodo AAF Adduct[38]

When AAF is modified at C7 with iodine in place of hydrogen, solution studies indicate a conformational effect different from base displacement in random sequence DNA. Only one to two basepairs are ruptured and little, if any, kinking or carcinogen-base stacking is observed.[39] It is the large Van der Waals radius of iodine that is responsible for the difference. Our calculations for the iodo derivative linked to dCpdG and polymer model building suggest a structure consistent with this data. The AAF is partly inserted into a B helix, with its large peripheral iodine atom protruding to the helix exterior. The guanine is syn at the modification site, and two base pairs are ruptured so that the AAF is not actually stacked with any adjacent bases. This model differs from earlier proposals suggesting that the guanine remains anti and the fluorene ring resides at the helix exterior.[39] Our calculations indicate that the guanine anti conformation is disfavored for the iodo derivative just as it is for the AAF adduct itself. The bulky acetyl group is responsible for the preference for syn-guanines in both cases.

2. The Guanine N^2 Adduct

a. Experimental Findings

While the majority of AAF adducts are linked to guanine C8 both in vitro and in vivo, a minor adduct to guanine N^2 has also been observed.[26] Little is known about the conformational or biological effects of this adduct, first because it is a minor constituent whose properties are overwhelmed by the major C8 adduct and second because of difficulty associated with its synthesis in DNA subunits. In mammalian systems the N^2 adduct is persistent[26] and it causes little DNA denaturation in solution.[40] Models consistent with this data, placing the AAF in the minor groove of an undeformed B helix[26,41] or in the major groove with some local conformational distortion[42] have been offered. Covalent intercalation models for

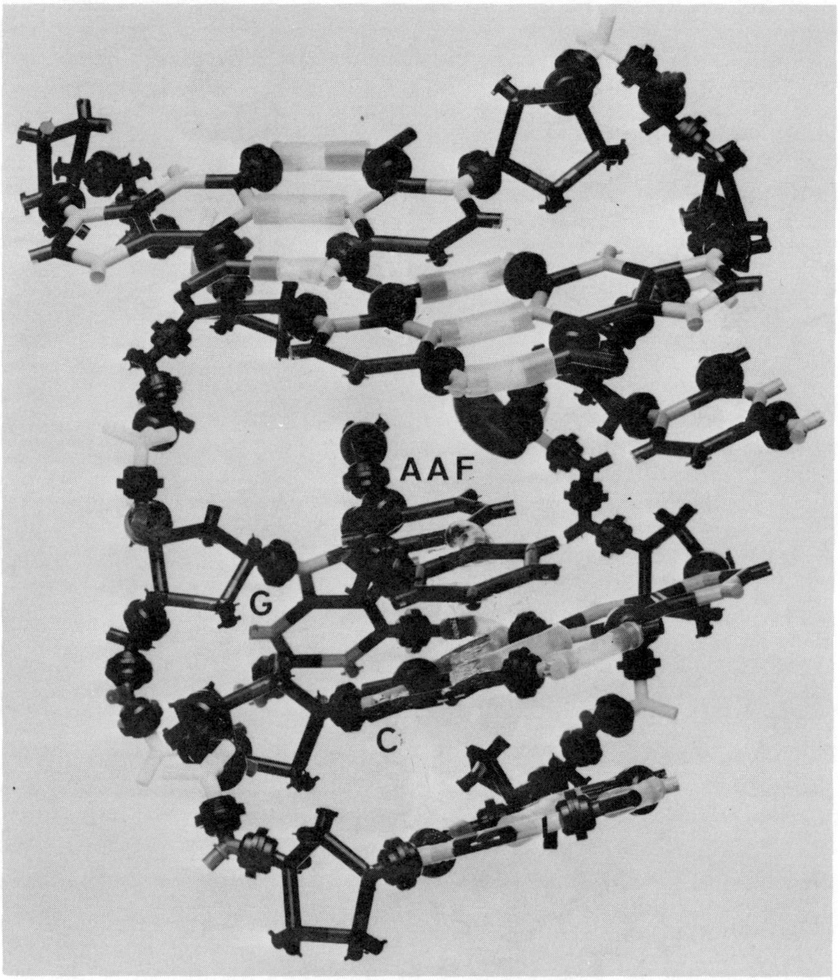

FIGURE 9. Base displacement in AAF modified Z-DNA: conformer of Figure 2b incorporated into a modified Z-form helix. (From Broyde, S. and Hingerty, B., *Carcinogenesis,* 6, 151, 1985. With permission.)

guanine N^2-linked polycyclic aromatic carcinogens, which feature carcinogen-base stacking have also been proposed.[43]

b. Computed Structures for the dCpdG Adduct and Polymer Model Building[44]

The global minimum for the guanine N^2-AAF adduct is shown in Figure 10. Interestingly, it has all the conformational features of the dCpdG residue of Z-DNA.[15] Yet, an attempt to build this conformer into a Z helix produced steric clashes that could not be alleviated by small torsion angle adjustments. However, it is possible to construct an intriguing model in which this dimer minimum is incorporated into the Z form in one direction with a junction to the B form in the other (Figure 11). The AAF resides at the exterior of this structure, hanging down into the shallow groove of the Z region. All basepairs are intact, in conformity with the solution data.

B. 2-Aminofluorene (AF)

1. Experimental Findings

Current evidence suggests that the deacetylated aminofluorene adduct is a lesion that is important for carcinogenesis in mammalian systems. In mammals, the majority of guanine

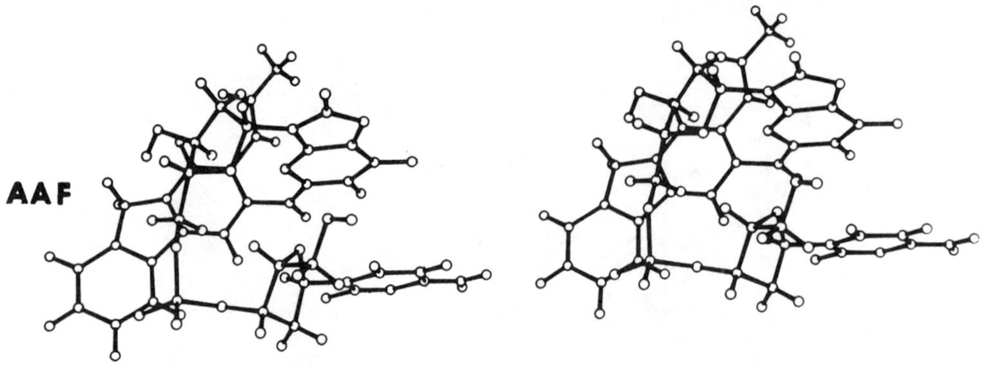

FIGURE 10. dCpdG modified at guanine N^2 in its Z-like lowest energy conformation, in stereo view. (From Hingerty, B. and Broyde, S., *Nucleic Acids Res.*, 11, 3241, 1983. With permission.)

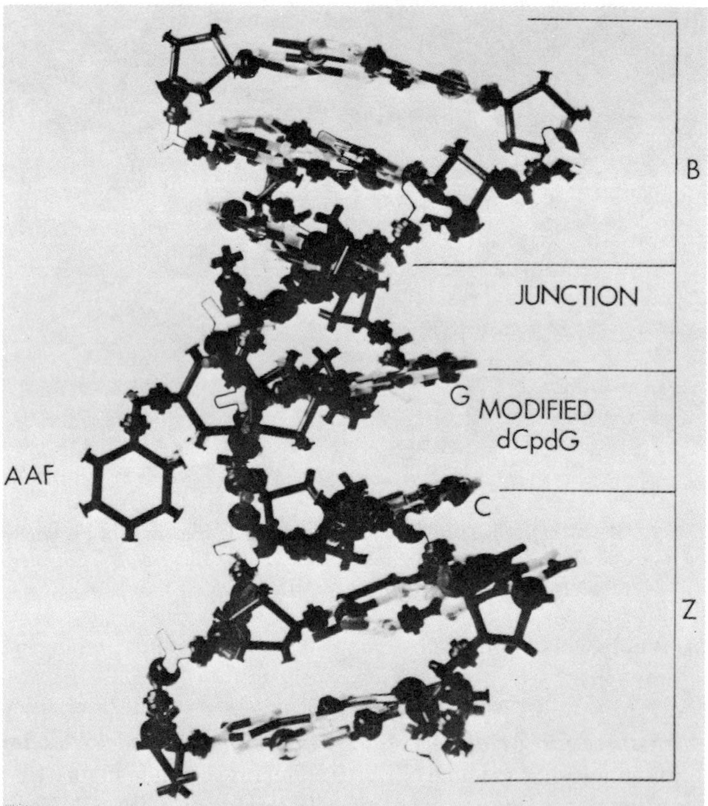

FIGURE 11. Conformer of Figure 10 incorporated into a B-Z junction. (From Hingerty, B. and Broyde, S., *Nucleic Acids Res.*, 11, 3241, 1983. With permission.)

C8 adducts are well known to be deacetylated.[24,25] Furthermore, in the *Salmonella typhymurium* microsome reversion assay for mutagenicity, administration of *N*-acetoxy-2-acetylaminofluorene in the presence of mammalian enzymes yields only deacetylated C8 adducts, and the mutation frequency correlates with the fraction of DNA that is modified.[45] In addition, the AF adduct is much more persistent than its acetylated counterpart in mammals.[46] In *E. coli* AF largely causes point mutations, while AAF mainly produces frameshifts.[11,16] Solution studies of this adduct suggest that it causes less severe structural alterations to the DNA

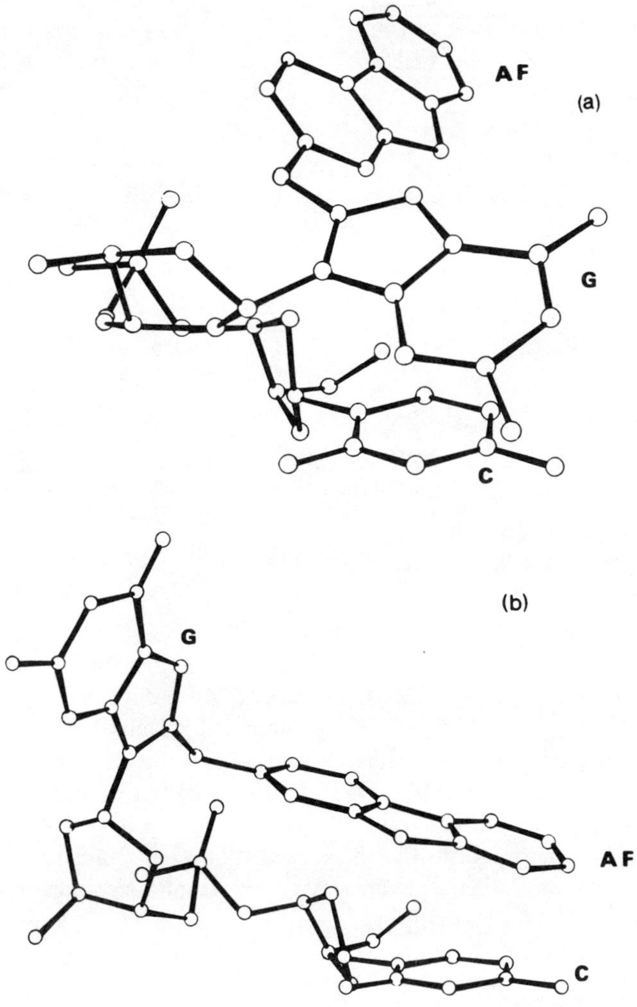

FIGURE 12. (a) A-B type base-base stacked conformation of dCpdG-AF; (b) perturbed B form carcinogen-base stacked conformation of dCpdG-AF. (From Broyde, S. and Hingerty, B., *Ann. N.Y. Acad. Sci.*, 435, 120, 1984. With permission.)

conformation than does AAF.[47-49] Evans et al. have presented a model that places the AF in the major groove of the B helix.[47]

2. Important Calculated Structures for the dCpdG Adduct and Polymer Model Building[50]

The low energy conformations of the AAF adduct are low energy in the deacetylated AF derivative as well. Indeed, the conformer shown in Figure 2a, which has carcinogen-base stacking with syn-guanine is also low energy for dCpdG-AF. However, replacement of the acetyl group with a hydrogen still has a profound conformational effect; the anti-domain of the guanine glycosidic torsion, which is strongly disfavored in the AAF adduct, is about equally preferred as syn in the AF adduct. Lipkowitz et al.[51] have also computed a greater preference for the anti-domain in AF vs. AAF modified deoxyguanosine.

Furthermore, we find two types of low energy guanine anti-conformations for AF, both of which are totally absent from the low energy domain of AAF. One type has base-base stacking and includes conformations like the normal A and B helixes (Figure 12a). A second

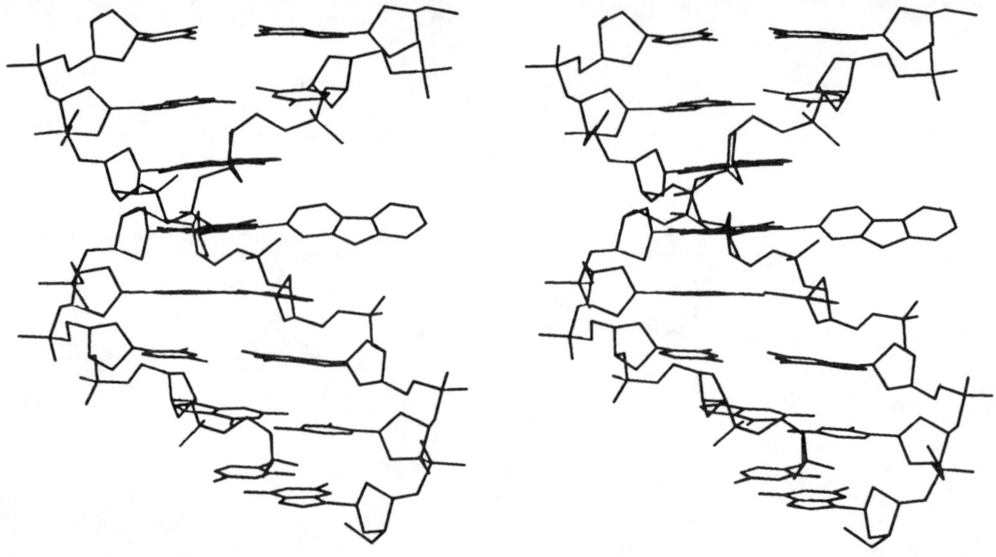

FIGURE 13. Carcinogen-base orientation of Figure 12a incorporated into a duplex B form followed by complete energy minimization, in stereo view.

type has carcinogen-base stacking. Indeed, a number of different DNA backbone conformers are capable of producing aminofluorene-base stacking with guanine both anti and syn. Among these is one conformer (Figure 12b) whose torsion angles are all in or near the A-B DNA range. A 30 to 40° rotation of the C5′–O5′ band from trans towards gauche minus is the largest distortion needed to produce this AF-base stacked state.

The existence of low energy forms like the normal A or B helix suggested the incorporation of these dimer minima into larger structures. An energy minimized duplex B form is shown in Figure 13. The AF causes virtually no distortion to the B-DNA duplex, residing in the major groove. An energy-refined single-stranded A form with AF in the groove is shown in Figure 14. The A form appears to be slightly preferred over the B in alternating G-C single strand but not in duplexes. The carcinogen-base stacked dimer conformation of Figure 12b, incorporated into a single stranded B form and completely energy refined, is shown in Figure 15. The possible significance of such a carcinogen-base stacked state will be discussed below for the 4-aminobiphenyl modified DNA.

C. 4-Aminobiphenyl (ABP)

1. Experimental Findings

While the biochemical activation, mutagenicity, and carcinogenicity of this chemical have been carefully studied in recent years,[45a] there does not appear to be any solution data elucidating its influence on DNA conformation, except for the work from our group and one other. It has been determined that a major adduct is formed to guanine C8 both in vitro and in vivo (in mammals). It is also the major adduct formed in the *Salmonella* assay. In addition, the number of C8 adducts correlates linearly with the number of revertants, just as for the AF adduct. Interestingly, the ABP guanine C8 adduct is far less mutagenic than the AF one in *Salmonella* strain 1538, although the two chemicals differ only in that the methylene bridge connecting the two phenyl rings is present in AF and absent in ABP. The C8-ABP adduct persists in rats to at least 29 days.[53] Lasko et al.[45b] have synthesized d(TGCA), modified at guanine C8 with ABP. The tetramer has been incorporated in *E. coli*, its mutation spectrum determined, and NMR data have been obtained.

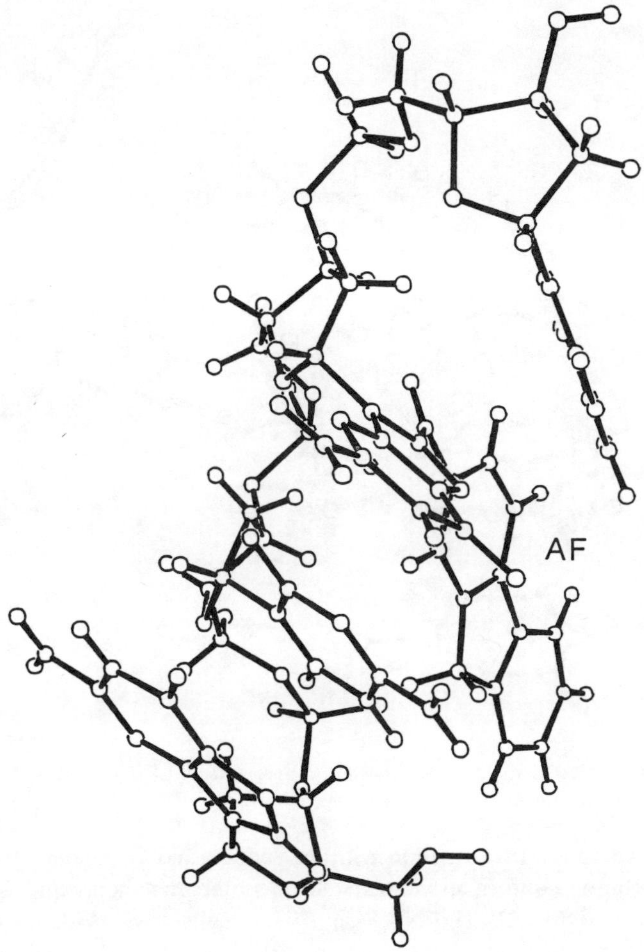

FIGURE 14. Conformer of Figure 12a incorporated in an A form followed by complete energy minimization.

2. Important Conformers for the Guanine C8 Adduct of dCpdG with ABP: Calculations and Synthesis

In order to obtain a molecular view of the ABP adduct to guanine C8 we first studied the dCpdG adduct. At this point we obtained access to supercomputing resources which made it possible to search the conformation space more extensively, with 4000 trials per adduct rather than the 500 we had been limited to previously. The three most important types of conformers have ABP-base stacking with anti-guanine (or high anti) (Figure 16), ABP-base stacking with syn-guanine (Figure 17), and base-base stacking with anti-guanine[54] (Figure 18). All are expected to make a contribution to the population distribution in solution. It is also worth noting that the phenyl rings are twisted with respect to one another in the low energy conformations, as has been observed for biphenyl in solution.[55] This may have a bearing on the diminished mutagenicity of ABP vis-à-vis AF[45] (see below). Low energy forms of particular interest to larger polymers are the B- and Z-type forms of Figure 18, and the perturbed B- and Z-like conformers of Figure 16, which have carcinogen-base stacking. In the latter, the B form is altered in that the C5′–O5′ bond is rotated from trans towards gauche minus, while the Z form is changed in that the guanine glycosidic torsion is rotated from syn to anti. To test the predictions, the dCpdG-ABP adduct was also syn-

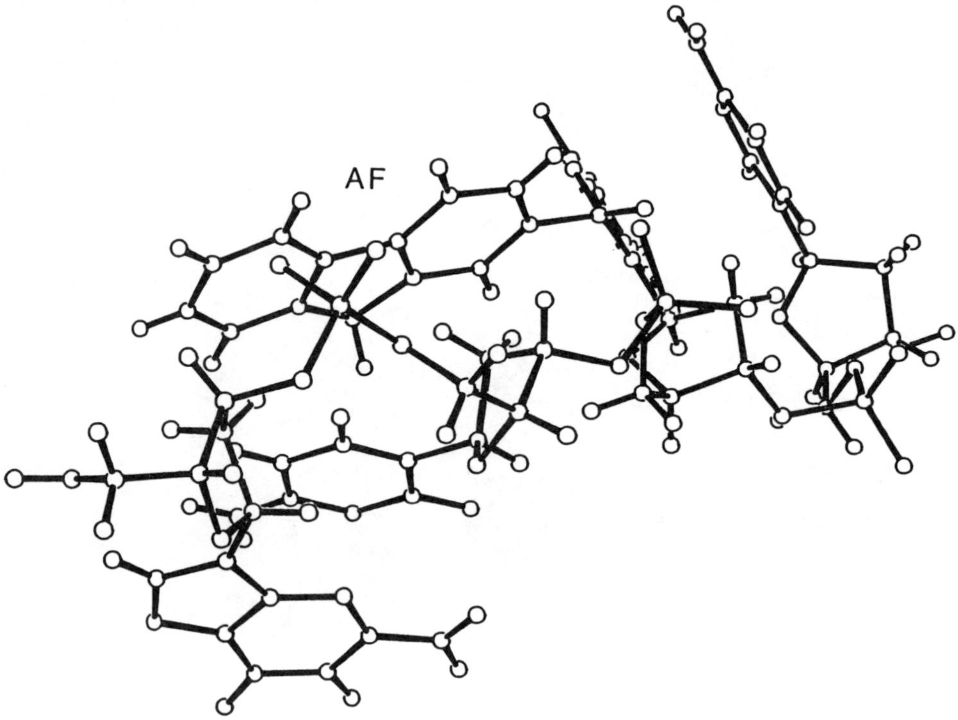

FIGURE 15. Conformer of Figure 12b incorporated in a single stranded B form followed by complete energy minimization.

thesized in a collaborative investigation with R. Shapiro and G. Underwood of New York University. Its solution conformation was studied by circular dichroism (CD) and nuclear magnetic resonance (NMR).[54] The NMR and CD results gave evidence of a significant contribution to the rotamer population mix by ABP-base stacked states, in which the innermost phenyl ring is stacked with cytidine as shown in Figure 17.

3. Polymer Models

In collaboration with us, A. R. Srinivasan has docked the conformers most relevant to the B and Z forms into the appropriate larger polymers to deduce the effect that ABP will have on these DNA conformations.[56] The A-B type conformer of Figure 18a incorporated in a duplex B helix is shown in Figure 19. The carcinogen is situated in the major groove and does not alter the double stranded structure in any way. On the other hand, the perturbed B-like form with carcinogen-base stacking would obviously distort the B form. When docked into a single stranded B helix (Figure 20) it causes a bend. The possible significance to mutagenesis of such a carcinogen-base stacked single stranded structure is discussed below.

The Z-like conformers (Figures 18b and 16b) have been incorporated into a Z helix as shown in Figure 21a and b. Both an undistorted Z helix with ABP at the helix exterior, and a pertubed Z form with carcinogen-base stacking and anti guanine at the modification site are seen to be feasible.

VI. CONFORMATION AND BIOLOGICAL PROPERTIES

Once a particular carcinogen-DNA adduct is formed, it must avoid error-free repair till a round of replication, if it is to be implicated in the initiation of cancer. It was Kriek[26] who first suggested that the molecular basis for selectivity in repair may be conformation de-

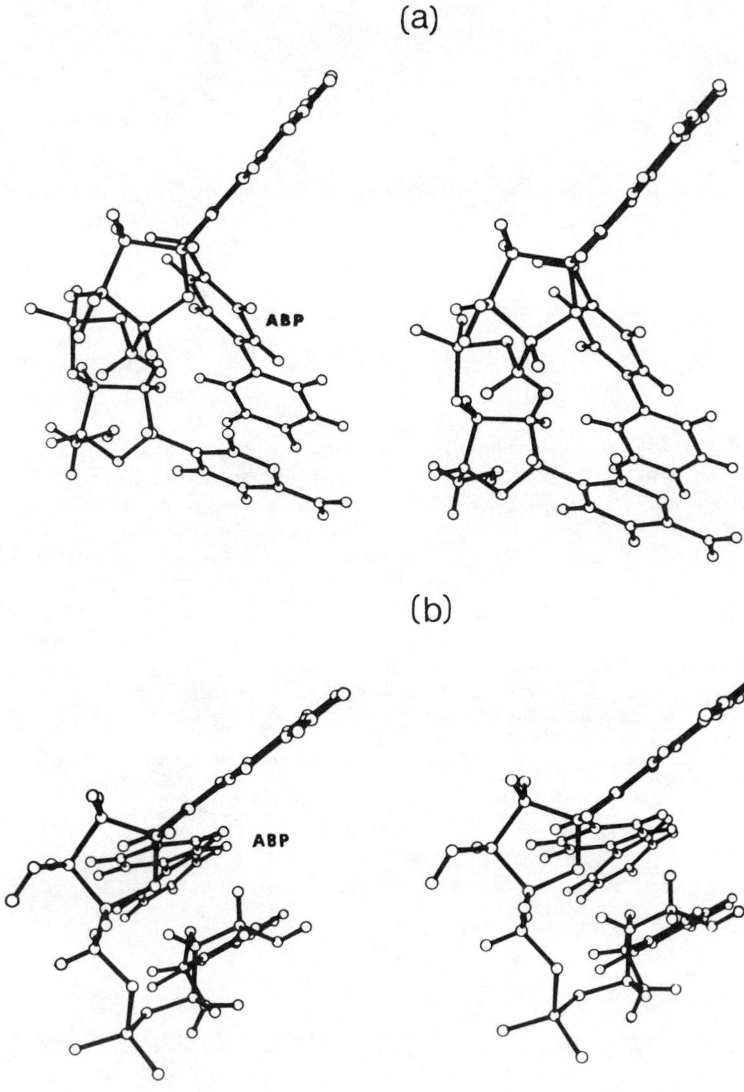

FIGURE 16. dCpdG-ABP minimum energy conformations with carcinogen-base stacking and guanine anti or high anti, in stereo view. (a) is a perturbed B form and (b) is a perturbed Z form.

pendent, and that nondistorting lesions might be the ones to persist. Our polymer models for A-B type helixes have suggested a hypothesis relating conformation, repair, and mutagenicity.[56] The calculations and polymer models for the persistent guanine C8-AF and ABP adducts reveal that the carcinogens can reside in the groove of an undeformed A or B helix, possibly escaping repair altogether because of their buried location in an undistorted DNA polymer. On the other hand, the AAF-C8 adduct, which is largely repaired in mammalian systems, has no low energy forms that are undistorted A-B type helixes. Instead, solution studies and model building indicate that it causes denaturation and carcinogen-base stacking in random sequence DNA. This severe perturbation might then invoke repair enzymes.

Those adducts that escape error-free repair have overcome a critical defense against the carcinogens. However, a further step is important, namely induction of a mutation at replication by the carcinogen-modified base. The mutagenic event actually takes place at the

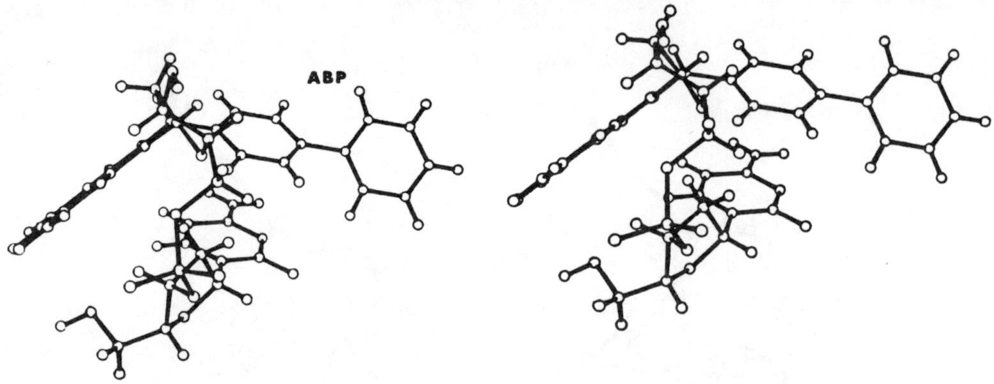

FIGURE 17. dCpdG-ABP global minimum energy conformation with carcinogen-base stacking and syn-guanine, in stereo view. (From Shapiro, R., Underwood, G., Zawadzka, H., Broyde, S., and Hingerty, B. E., *Biochemistry,* 25, 2198, 1986. With permission.)

(a)

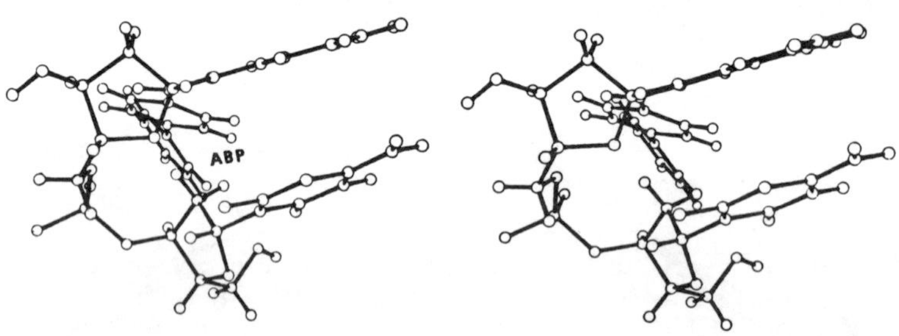

(b)

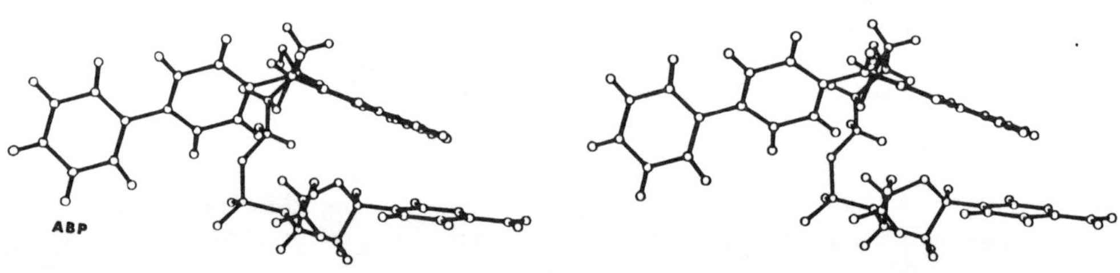

FIGURE 18. (a) dCpdG-ABP A-B type conformation with base-base stacking and anti-guanine, in stereo view; (b) dCpdG-ABP Z-type conformation, in stereo view.[56]

single-stranded replication fork, where the DNA is unwinding and hence conformationally mobile. We have suggested that a carcinogen-base stacked state like that shown in Figures 15 and 20 might be readily formed under these circumstances, and that such a conformation may be one to cause error by the replication machinery.[56] The concept that a covalently linked carcinogen-base stacked form might actually be a mutagenic conformation is analogous to the situation of mutagenic equilibrium-bound DNA-intercalating agents. It might further be speculated that the diminished mutagenicity in *Salmonella typhimurium* 1538 of ABP vs. AF may result from diminished stability of carcinogen-base stacked states, inherent in ABP because it lacks the central ring formed by the methylene bridge of the fluorene moiety, so that the two phenyl rings are twisted in solution. It is also noteworthy that a similar pair of

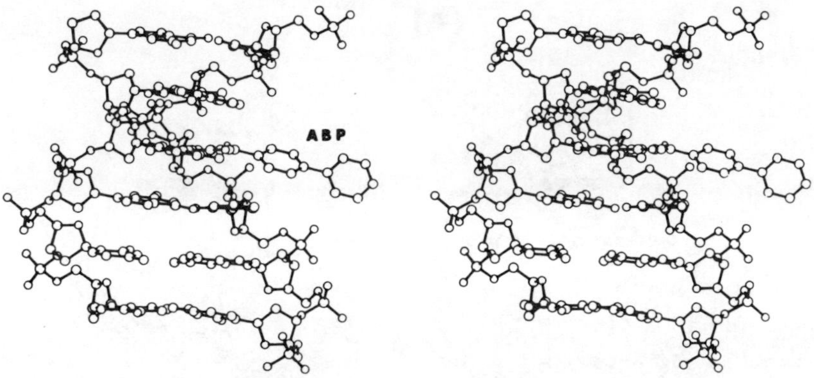

FIGURE 19. Carcinogen-base orientation of minimum energy conformation shown in Figure 18a docked into a duplex B helix, in stereo view. (From Broyde, S., Hingerty, B. E., and Srinivasan, A. R., *Carcinogenesis,* 6, 719, 1985. With permission.)

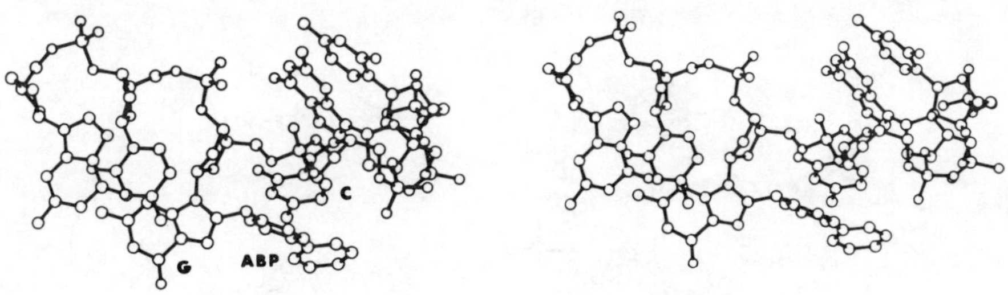

FIGURE 20. B-like carcinogen-base stacked conformer of Figure 16a docked into a single stranded B form, in stereo view. (From Broyde, S., Hingerty, B. E., and Srinivasan, A. R., *Carcinogenesis,* 6, 719, 1985. With permission.)

A-B type conformers, one with base-base stacking and a second with carcinogen-base stacking, has been computed for the dCpdG (+) anti-benzo(*a*)pyrene diolepoxide adduct to guanine N^2,[57] which has been identified as the tumorigenic lesion produced by the benzo(*a*)pyrenes.[58] According to our hypothesis then, both externally bound, undeformed states and carcinogen-base stacked states must be conformationally accessible for a particular carcinogen-base adduct to be mutagenic. Diminished accessibility to either state should lead to lowered mutagenic potential.

Our studies have also revealed some Z form models for carcinogen-modified DNAs. In guanine C8 adducts, these include both externally situated carcinogen orientations, as well as inserted forms. For the guanine N^2-AAF adduct, the calculations and model building suggest a B-Z junction model, with carcinogen at the junction in the Z direction groove. The observed persistence of this adduct[26] might again result from the buried location of the carcinogen at an undenatured duplex, or Z forms may be less prone to repair.[59] While the biological role of the left-handed Z form has not yet been defined, it has been suggested that it is involved in the control of gene expression.[60] One could imagine that induction of a Z form or a B-Z junction in a genetic control region containing alternating pyrimidine-purine sequences that are prone to adopt the Z form might alter the normal level of transcription in the adjacent coding region.

Oncogene activation may occur by mutation and/or by over-expression of the unmutated normal gene,[10] and current evidence indicates that at least two genetic events are needed to produce transformed cells from normal ones.[9,61] That chemical carcinogens can produce

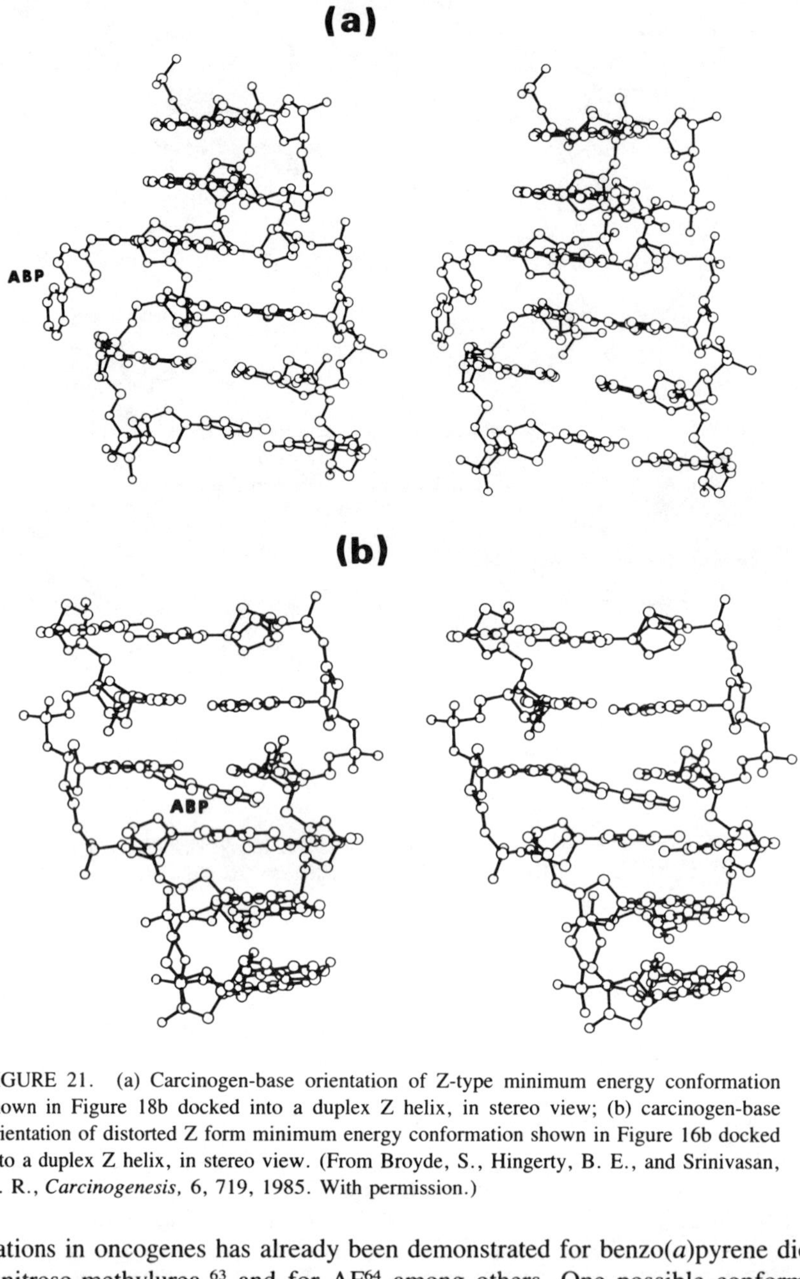

FIGURE 21. (a) Carcinogen-base orientation of Z-type minimum energy conformation shown in Figure 18b docked into a duplex Z helix, in stereo view; (b) carcinogen-base orientation of distorted Z form minimum energy conformation shown in Figure 16b docked into a duplex Z helix, in stereo view. (From Broyde, S., Hingerty, B. E., and Srinivasan, A. R., *Carcinogenesis,* 6, 719, 1985. With permission.)

point mutations in oncogenes has already been demonstrated for benzo(*a*)pyrene diolepoxide,[62] for nitroso-methylurea,[63] and for AF[64] among others. One possible conformational mechanism for producing such mutations, described above, has been suggested by our calculations and model building, although others such as depurination could certainly be important as well. Over-expression of the normal oncogene product induced by a chemical carcinogen that favors a DNA Z form in a carcinogen-modified control region, might be a plausible mechanism for carcinogen-induced deregulation, possibly mediated by genetic rearrangement.

ACKNOWLEDGMENT

This work was supported jointly by PHS Grant #1 RO1CA28038-06, awarded by the National Cancer Institute, DHHS, DOE Contract #DE-AC02-81 ERG0015, NSF Grant

#DMB8416009, and by the Office of Health and Environmental Research, U.S. Department of Energy, under Contract #DE-AC05-84OR 21400 with Martin-Marietta Energy Systems, Inc. (BH).

REFERENCES

1. **Rehn, L.,** Blasengeschwulste bei fuchsin-arbeitern, *Arch. Klin. Chir.,* 50, 588, 1895.
2. **Case, R., Hosker, M., McDonald, D., and Pearson, J.,** Tumors of the urinary bladder in workmen engaged in the manufacture and use of certain dye intermediates in the British Chemical Industry. I. The role of aniline, benzidine, 1-naphthylamine, and 2-naphthylamine, *Br. J. Ind. Med.,* 11, 75, 1954.
3. **Case, R. and Pearson, J.,** Further consideration of the role of aniline and of the manufacture of auramne and magenta(fuchsine) as possible causative agents, *Br. J. Ind. Med.,* 11, 213, 1954.
4. **Radomski, J., Deichmann, W., Altman, N., and Radomski, T.,** Failure of pure 1-napthylamine to induce bladder tumors in dogs, *Cancer Res.,* 40, 3537, 1980.
5. **National Cancer Institute**, Bioassay of aniline hydrochloride for possible carcinogenicity, *Natl. Cancer Inst. Carcinogen. Tech. Rep. Ser. 130,* NCI, Bethesda, Md., 1978, 1.
6. *IARC Monogr. Evaluation Carcinogenic Risk Chem. Man,* Vol. 1, IARC, Lyon, 1972, 74.
7. **Wilson, R., De Eds, F., and Cox, A.,** The toxicity and carcinogenic activity of 2-acetylaminofluorene, *Cancer Res.,* 1, 595, 1941.
8. **Garner, R., Martin, N., and Clayson, D.,** Carcinogenic aromatic amines and related compounds, *Chemical Carcinogens,* 2nd ed., Searle, C. E., Ed., ACS Monogr. 182, American Chemical Society, Washington, D.C., 1984, 175.
9. **Anon.** Chemical carcinogens; a review of the science and its associated principles, February 1985, *Fed. Reg.,* March 14, 10372, 1985.
10. **Weinberg, R.,** A molecular basis of cancer, *Sci. Am.,* November 1983, 126.

11a. **Srinivasan, A. R. and Olson, W.,** Polynucleotide conformations in real solutions — a preliminary theoretical estimate, *Fed. Proc.,* 39, 2199, 1980.

11b. **Olson, W.,** Spatial configuration of ordered polynucleotide chains. V. Conformational energy estimates of helical structure, *Biopolymers,* 17, 1015, 1978.

12. **Hingerty, B. and Broyde, S.,** Conformation of the deoxydinucleoside monophosphate dCpdG modified at carbon 8 of guanine with 2-(acetylamino)fluorene, *Biochemistry,* 21, 3243, 1982.
13. **Powell, M.,** An efficient method for finding the minimum of a function of several variables without calculating derivatives, *Computer J.,* 7, 155, 1964.
14. **Saenger, W.,** *Principles of Nucleic Acid Structure,* Springer-Verlag, New York, 1984.
15. **Wang, A., Quigley, G., Kolpak, F., Van der Marel, G., van Boom, J., and Rich, A.,** Left handed double helical DNA: variations in the backbone conformation, *Science,* 211, 171, 1980.
16. **Fuchs, R. P. P.,** DNA binding spectrum of the carcinogen N-acetoxy-N-2-acetylaminofluorene significantly differs from the mutation spectrum, *J. Mol. Biol.,* 177, 173, 1983.
17. **Bichara, M. and Fuchs, R. P. P.,** DNA binding and mutation spectra of the carcinogen N-2-aminofluorene in *E. coli:* a correlation between the conformation of the premutagenic lesion and mutation specificity, *J. Mol. Biol.,* 183, 341, 1985.
18. **Rao, S., Keepers, J., and Kollman, P.,** The structure of d(CGCGAAT TCGCG). d(CGCGAATTCGCG): the incorporation of a thymine photodimer into a B-DNA helix, *Nucleic Acids Res.,* 12, 4789, 1984.
19. **Pearlman, D., Holbrook, S., and Kim, S. H.,** Molecular models for DNA damaged by photoreaction, *Science,* 227, 1304, 1985.
20. **Stellman, S., Hingerty, B., Broyde, S., Subramanian, E., Sato, T., and Langridge, R.,** Structure of guanosine-3′,5′-cytidine monophosphate. I. Semi-empirical potential energy calculations and model building, *Biopolymers,* 12, 2731, 1973.
21. **Broyde, S., Wartell, R., Stellman, S., Hingerty, B., and Langridge, R.,** Classical potential energy calculations for ApA, CpC, GpG, and UpU. The influence of the bases on RNA subunit conformations, *Biopolymers,* 14, 1597, 1975.
22. **Hingerty, B., Brown, R., and Jack, A.,** Further refinement of the structure of yeast $tRNA^{phe}$, *J. Mol. Biol.,* 124, 523, 1978.
23. **Broyde, S. and Hingerty, B.,** The influence of terminal 3′,5′, phosphates on conformations of dApdA, *Nucleic Acids Res.,* 6, 2165, 1979.
24. **Irving, C. and Veazey, R.,** Persistent binding of 2-acetylaminofluorene to rat liver DNA in vivo and consideration of the mechanism of binding of N-hydroxy-2-acetylaminofluorene to rat liver nucleic acids, *Cancer Res.,* 29, 1799, 1969.

25. **Allaben, W., Weis, C., Fullerton, N., and Beland, F.,** Formation and persistence of DNA adducts from the carcinogen N-hydroxy-2-acetylaminofluorene in rat mammary gland *in vivo, Carcinogenesis,* 4, 1067, 1983.
26. **Kriek, E.,** Carcinogenesis by aromatic amines, *Biochim. Biophys. Acta,* 355, 177, 1974.
27. **Visser, A. and Westra, J.,** Partial persistency of 2-aminofluorene and N-acetyl-2-aminofluorene in rat liver DNA, *Carcinogenesis,* 2, 737, 1981.
28. **Grunberger, D. and Weinstein, I.,** Conformational changes in nucleic acids modified by chemical carcinogens, *Chemical Carcinogens and DNA,* Grover, P. L., Ed., CRC Press, Boca Raton, Fla., 1979, 59.
29. **Santella, R. and Grunberger, D.,** Induction of the base displacement or Z conformation in DNA by N-2-acetylaminofluorene modification, *Environ. Health Perspect.,* 49, 107, 1983.
30. **Fuchs, R. P. P. and Daune, M.,** Physical studies on deoxyribonucleic acid after covalent binding of a carcinogen, *Biochemistry,* 11, 2659, 1972.
31. **Fuchs, R. P. P. and Daune, M.,** Dynamic structure of DNA modified with the carcinogen N-acetoxy-N-2-acetylaminofluorene, *Biochemistry,* 13, 4435, 1974.
32. **Fuchs, R. P. P., Lefevre, J., Pouyet, J., and Daune, M.,** Comparative orientation of the fluorene residue in native DNA modified by N-acetoxy-N-2-acetylaminofluorene and two 7-halogeno derivatives, *Biochemistry,* 15, 3347, 1976.
33. **Hingerty, B. and Broyde, S.,** Conformation of the deoxydinucleoside monophosphate dCpdG modified at carbon 8 of guanine with 2-(Acetylamino)fluorene, *Biochemistry,* 21, 3243, 1982.
34. **Olson, W.,** Understanding the motions of DNA, *Biomolecular Stereodynamics,* Vol. 1, Sarma, R., Ed., Adenine Press, Albany, N.Y., 1981, 327.
35. **Neidle, S., Kuroda, R., Broyde, S., Hingerty, B. E., Levine, R., Miller, D., and Evans, F.,** Studies on the conformation and dynamics of the C8-substituted guanine adduct of the carcinogen acetylaminofluorene; model for a possible Z-DNA modified structure, *Nucleic Acids Res.,* 12, 8219, 1984.
36. **Hanau, L., Santella, R., Grunberger, D., and Erlanger, B.,** An immunochemical examination of acetylaminofluorene-modified poly(dG-dC).poly(dG-dC) in the Z conformation, *J. Biol. Chem.,* 259, 173, 1984.
37. **Broyde, S. and Hingerty, B.,** Base displacement in AAF-modified Z-DNA, *Carcinogenesis,* 6, 151, 1985.
38. **Broyde, S. and Hingerty, B.,** An internal fluorene model for iodo-N-2-acetylaminofluorene modified DNA, *Chem. Biol. Interact.,* 47, 69, 1983.
39. **Fuchs, R. P. P., Lefevre, J., Pouget, J., and Daune, M.,** Comparative orientation of fluorene residue in native DNA modified by N-acetoxy-N-2-acetyl-aminofluorene and 2 7-halogen-o derivatives, *Biochemistry,* 15, 3347, 1976.
40. **Yamasaki, H., Pulkrabek, P., Grunberger, D., and Weinstein, I.,** Differential excision from DNA of the C-8 and N2 guanosine adducts of N-acetyl-2-aminofluorene by single strand-specific endonucleases, *Cancer Res.,* 37, 3756, 1977.
41. **Beland, F.,** Computer generated graphic models of the N2-substituted deoxyguanosine adducts of 2-acetylaminofluorene and benzo(a)pyrene and the O6-substituted deoxyguanosine adduct of 1-naphthylamine in the DNA double helix, *Chem. Biol. Interact.,* 22, 329, 1978.
42. **Kadlubar, F.,** A transversion mutation hypothesis for chemical carcinogenesis by N2-substitution of guanine in DNA, *Chem. Biol. Interact.,* 31, 255, 1980.
43. **Drinkwater, N., Miller, J., Miller, E., and Yang, N.,** Covalent intercalative binding to DNA in relation to the mutagenicity of hydrocarbon epoxides and N-acetoxy-2-acetylaminofluorene, *Cancer Res.,* 38, 3247, 1978.
44. **Hingerty, B. and Broyde, S.,** AAF linked to the guanine amino group: a B-Z junction, *Nucleic Acids Res.,* 11, 3241, 1983.

45a. **Beland, F., Beranek, D., Dooley, K., Heclich, R., and Kadlubar, F.,** Arylamine-DNA adducts *in vitro* and *in vivo:* their role in bacterial mutagenesis and urinary bladder carcinogenesis, *Environ. Health Perspect.,* 49, 125, 1983.

45b. **Lasko, D., Weis, C., DeRosa, L., Essigmann, J., and Kadlubar, F.,** Synthesis of a tetranucleotide containing the major DNA adduct of 4-aminobiphenyl, *Proc. 75th Annu. Meet. Am. Assoc. Cancer Res.,* Vol. 25, Am. Assoc. Cancer Res., Philadelphia, 1984, 90.

46. **Poirer, M., True, B'Ann, and Laishes, B.,** Determination of 2-acetylaminofluorene adducts by immunoassay, *Environ. Health Perspect.,* 49, 93, 1982.
47. **Evans, F., Miller, D., and Beland, F.,** Sensitivity of the conformation of deoxyguanosine to binding at the C-8 position by N-acetylated 2-aminofluorene, *Carcinogenesis,* 1, 955, 1980.
48. **Leng, M., Ptak, M., and Rio, P.,** Conformation of acetylaminofluorene and aminofluorene modified guanosine and guanosine derivatives, *Biochem. Biophys. Res. Commun.,* 96, 1095, 1980.
49. **Santella, R., Kriek, E., and Grunberger, D.,** Circular dichroism and proton magnetic resonance studies of dApdG modified with 2-aminofluorene and 2-acetylaminofluorene, *Carcinogenesis,* 1, 897, 1980.
50. **Broyde, S. and Hingerty, B.,** Conformation of 2-aminofluorene-modified DNA, *Biopolymers,* 22, 2423, 1983.

51. **Lipkowitz, K., Chevalier, T., Widdifield, M., and Beland, F.,** Force field conformational analysis of aminofluorene and acetylaminofluorene substituted deoxyguanosine, *Chem. Biol. Interact.*, 40, 57, 1982.
52. **Conner, B., Yoon, C., Dickerson, J., and Dickerson, R.,** Helix geometry and hydration in an A-DNA tetramer:[I]C-C-G-G, *J. Mol. Biol.*, 174, 663, 1984.
53. **Gupta, R. and Dighe, N.,** Formation and removal of DNA adducts in rat liver treated with N-hydroxy derivatives of 2-acetylaminofluorene, 4-acetylaminobiphenyl, and 2-acetylaminophenanthrene, *Carcinogenesis,* 5, 343, 1984.
54. **Shapiro, R., Underwood, G., Zawadzka, H., Broyde, S., and Hingerty, B. E.,** Conformation of d(CpG) modified by the carcinogen 4-aminobiphenyl: a combined experimental and theoretical analysis, *Biochemistry,* in press.
55. **Canselier, J. and Cassoux, P.,** An unusual application of the Faraday effect: conformational analysis in bridged biphenylic systems, *J. Mol. Struct.*, 39, 301, 1977.
56. **Broyde, S., Hingerty, B. E., and Srinivasan, A. R.,** Influence of the carcinogen 4-aminobiphenyl on DNA conformation, *Carcinogenesis,* 6, 719, 1985.
56a. **Broyde, S. and Hingerty, B. E.,** Energy minimized structures of carcinogen-DNA adducts: 2-acetylaminofluorine and 2-amino-fluorene, *J. Biomolec. Struct. Dynam.*, 4, 365, 1986.
57. **Hingerty, B. and Broyde, S.,** A conformational analysis of the (+) anti BPDE adduct to the guanine amino group of dCpdG, *J. Biomol. Struct. Dyn.*, 1, 905, 1983.
58. **Conney, A. H.,** Induction of microsomal enzymes by foreign chemicals and carcinogenesis by polycyclic aromatic hydrocarbons, *Cancer Res.*, 42, 4875, 1982.
59. **Boiteux, S., de Oliveira, R., and Laval, J.,** The *Escherichia coli* O^6-methylguanine-DNA methyltransferase does not repair O^6-methylguanine residues when present in Z-DNA, *J. Biol. Chem.*, 260, 8711, 1985.
60. **Rich, A., Nordheim, A., and Azorin, F.,** Stabilization and detection of natural left-handed Z-DNA, *J. Biomol. Struct. Dyn.*, 1, 1, 1983.
61. **Land H., Parada, L., and Weinberg, R.,** Tumorigenic conversion of primary embryo fibroblasts requires at least two cooperating oncogenes, *Nature (London),* 304, 596, 1983.
62. **Marshall, C., Vousden, K., and Phillips, D.,** Activation of c-Ha-ras-1 proto-oncogene by *in vitro* modification with a chemical carcinogen, benzo(a)pyrene diol-epoxide, *Nature (London),* 310, 586, 1984.
63. **Sukumar, S., Notario, V., Martin-Zanca, D., and Barbacid, M.,** Induction of mammary carcinomas in rats by nitroso-methylurea involves malignant activation of H-ras-1 locus by single point mutations, *Nature (London),* 306, 658, 1983.
64. **Wiseman, R., Stowers, S., Miller, E., Anderson M., and Miller, J.,** Activating mutations of the c-Ha-*ras* proto-oncogene in chemically induced hepatomas of the male B6C3F1 mouse, *Proc. Natl. Acad. Sci. U.S.A.*, 83, 5825, 1986.

Chapter 6

CRYSTALLOGRAPHIC AND COMPUTER-MODELING STUDIES ON THE METABOLITES OF POLYCYCLIC AROMATIC HYDROCARBONS AND THEIR INTERACTIONS WITH NUCLEIC ACIDS

R. Kuroda and S. Neidle

TABLE OF CONTENTS

I. INTRODUCTION

The induction of malignant disease by means of an environmental chemical was first noted in 1775 by Percival Pott, who found a high incidence of scrotal cancer among chimney sweeps. Subsequent reports in the 19th century repeatedly related exposures to materials such as coal tar, lubricating oils, and aniline dyes, to elevated incidence of a number of cancers. The identification of active constituents of coal-tar as polycyclic aromatic hydrocarbons (PAH) in the 1930s by Kennaway, Cook and colleagues,[1,2] led directly to the modern study of chemical carcinogenesis.[3] Subsequent findings that PAH are ubiquitous pollutants of modern industrial society have provided added impetus.[4] The importance and relevance of the subject have and continue to result in many reviews and monographs; in this chapter, we shall refer to those most relevant to the theme of understanding at the molecular level, structure-activity relationships between these molecules and their important biological receptor, DNA.[5-9] The biological chemistry and molecular biology of polycyclic aromatics, especially benzo(*a*)pyrene (BP), have produced a vast literature. Accordingly, no attempt will be made to review these areas in a comprehensive manner. The overall emphasis of this chapter is the application of crystallographic and computational methods to aspects of polycyclic hydrocarbon action and the interplay between them.

A. Metabolism and Reactivity

Different polycyclic hydrocarbons can show orders of magnitude difference in carcinogenic activity (Figure 1). Examination of the hydrocarbon chemical structures alone has not been able to satisfactorily explain these differences, even when electronic structure factors were invoked. The identification of specific regions in a polycyclic hydrocarbon (Figure 2) that have particular electronic properties, especially a reactive K region, enabled carcinogenic ranking orders to be correctly predicted in some instances;[10] there are, however, many exceptions to such rules, which suggest that the hydrocarbons themselves are not ultimate carcinogenic agents.

The central role of metabolic activation was first realized following the finding that upon applying BP and other polycyclics to mouse skin, a small fraction becomes covalently bound, and that only the extent of binding to cellular DNA correlates with carcinogenic potency.[11] Initially, it was believed that K-region epoxides were the responsible compound;[12] subsequent studies showed that the reactive compounds are diol epoxides, with the epoxide group in the bay region of the hydrocarbon.[13-18] In the case of BP, initial metabolism is to the 7,8-epoxide, then to the 7,8-diol, and finally to the 7,8-diol-9,10-epoxide (BPDE). These processes (Figure 3) are at least initially catalyzed by cytochrome P450 enzymes (the mixed-function oxygenases), with ring opening of the 7,8-epoxide catalyzed by the epoxide hydrolase enzyme. The highly reactive electrophile diol epoxide this produces can then form a carbonium ion. This can directly react with cellular macromolecules such as DNA, become involved in detoxification pathways, or be hydrolyzed to the 7,8,9,10-tetrol. Bay-region reactivities have received quantum-mechanical rationalizations based on estimates of the relative increase in delocalization energy on proceeding from a diol epoxide to the reactive cation,[20,21] which indicate that the stability of this cation can correlate with carcinogenicity, and that nonbay-region cations would be both less stable and less reactive. The bay-region theory[21] works well for a given hydrocarbon (especially for BP, on which much of these theoretical studies have been performed), but rather less well for comparisons between hydrocarbons. This is not surprising since the theory takes no account of other major factors involved in biological response. Particularly relevant to this review are interactions with DNA; stereochemical and electronic aspects of these would need to be taken into account for a more complete bay region theory. There is now increasing evidence that the critical carbonium ion is generated from the diol epoxide when the latter is bound to DNA in an

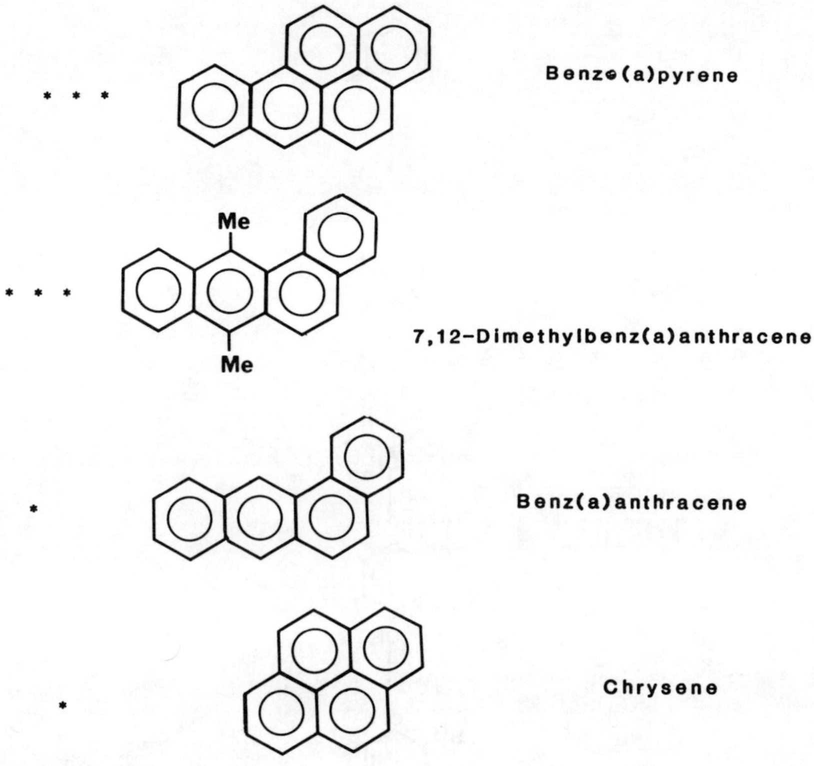

FIGURE 1. Some representative polycyclic aromatic hydrocarbons. Their relative carcinogenicities are indicated by asterisks.

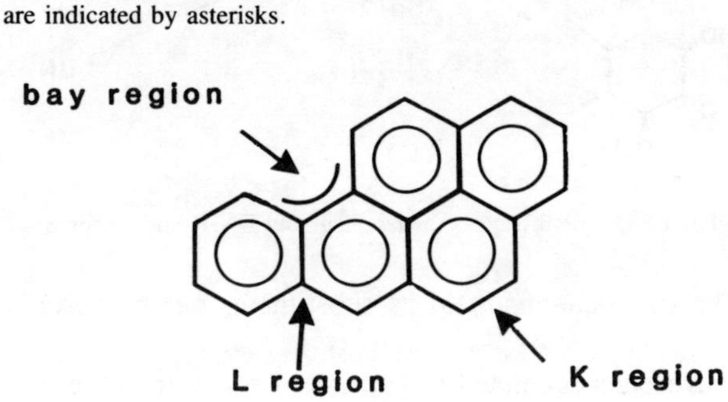

FIGURE 2. The reactive regions in a PAH.

initial recognition mode. Charge density and delocalization energy calculations have yet to be performed for carbonium ions of PAH in such an environment.

B. Stereochemical Aspects

The enzymatic metabolism of polycyclics is highly stereospecific. For BP the pathway indicated in Figure 3 involves the (+)-7,8-epoxide initially being formed, then *trans* ring-opening to the (+)-7,8-diol. The principal product of further oxidation is the (+)-anti isomer of the 7,8-diol 9,10-epoxide.[22-24] This is one of four possible stereoisomers (Figure 4). Much biological work has been performed on anti and syn forms. In general, the (±)-anti is the more biologically active by far compared to the (±)-syn in terms of tumor-initiating and mutagenic properties,[15,25] and the resolved (+)-anti-enantiomer has very much higher tumorigenicity than the (−) form. It is generally accepted that this (+) form is the ultimate

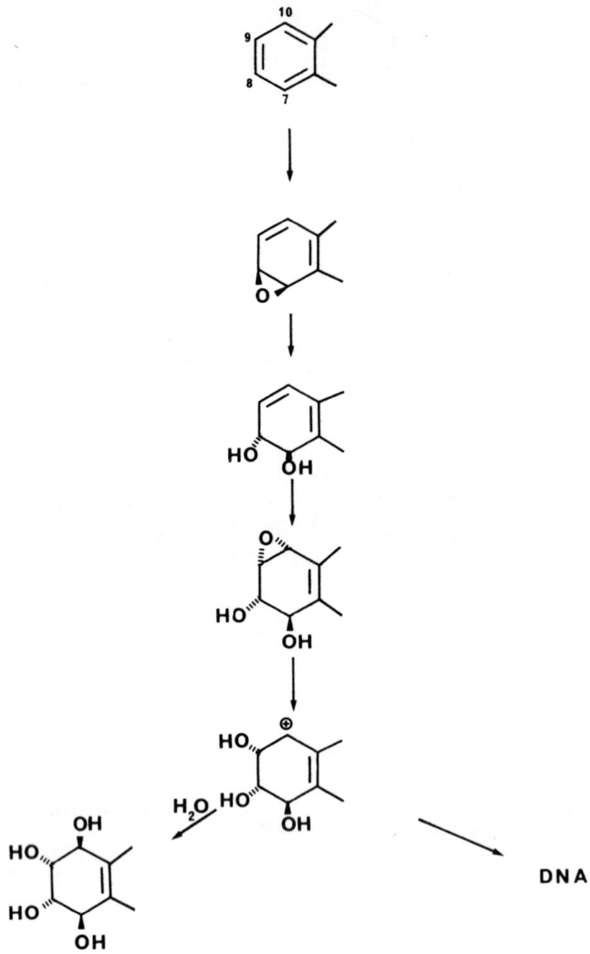

FIGURE 3. Pathways of metabolism and DNA binding for BP.

metabolite of BP. The stereochemistry of its substituents can be described as 7β, 8α, 9α, 10α.[27]

The syn isomers have been predicted to be more reactive than the anti ones on the basis of stereochemical considerations;[19] this has been borne out by studies on the rate of nucleophilic attack on the epoxide ring in solution[28,29] in vitro. The syn isomers are also mutagenic, to a greater extent than the anti in *Salmonella typhimurium*,[15,30] although the reverse is true in mammalian cells. The syn isomers are only very weakly carcinogenic.[26]

Numerous studies on PAH such as benz(*a*)anthracene (BA), 7,12-dimethylbenz(*a*)anthracene, and chrysene have similarly shown the critical importance of metabolic activation and stereochemistry of resulting metabolites in relation to biological activity.[7,12,31] X-ray crystallographic studies have been reported on some of these metabolites as well as on a number of BP derivatives, and are discussed below. By contrast, in vitro studies of interactions with DNA have almost exclusively concentrated on BP metabolites as the archetypal chemical carcinogens. This has meant that an understanding at the DNA interaction level, of differences in behavior between polycyclics with widely varying carcinogenicities has yet to be achieved.

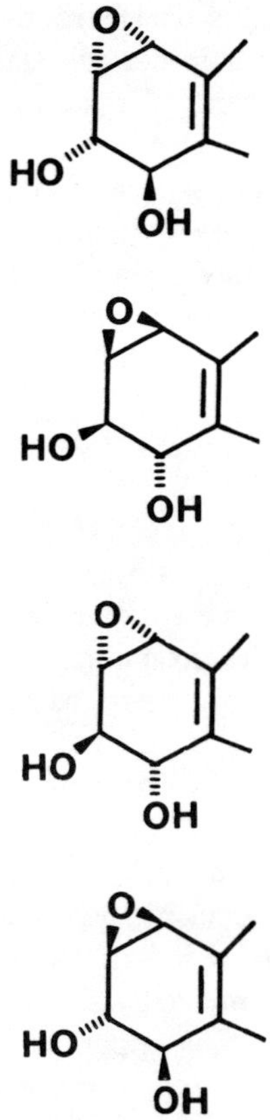

(+)- anti

(−)-anti

(+)-syn

(−)-syn

FIGURE 4. The stereoisomers of BP 7,8-diol-9,10-epoxide.

II. CRYSTALLOGRAPHIC STUDIES

Overview

X-ray crystallography has long been established as the most powerful technique for the determination of detailed molecular structure. The advent of increasingly sophisticated nuclear magnetic resonance (NMR) methodology has not as yet seriously challenged this preeminence. Rather, NMR provides information on more dynamic features of molecular structure in solution; the assignment of total stereochemistry for complex molecules (such as oligonucleotides or proteins) is currently limited to a maximum molecular weight of approximately 10 kdaltons, and generally does not have the certainty and richness of detail uniquely displayed by the X-ray method. The techniques of X-ray crystallography have been well described in a number of monographs[32-35] and will not be presented here. However, it is important for noncrystallographers to realize the accuracy and limitations of the method in relation to the specific structure determinations discussed in this review, and so these topics are discussed below. An X-ray crystallographic analysis of molecule (or in some instances an assembly of molecules), results in the knowledge of the three-dimensional spatial arrangement of the molecular species. This information is available as a set of coordinates, with reference to the axial system of the crystallographic unit cell. Standard geometric procedures are used to translate this information into the derived structural data of bond lengths and angles, bond torsion angles and least-squares planes, as well as hydrogen bond and other nonbonded distances. In addition, since the X-ray experiment provides a time, and hence, spatially averaged view of the internal molecular arrangement inside a crystal, individual atomic thermal-motion parameters may be obtained from the experimental diffracted X-ray intensity data. The accuracy of this data is routinely examined by means of statistical analysis of the data itself, and results in derived standard deviations for the atomic coordinates and thermal parameters and hence for the parameters of bonding geometry. These estimated standard deviations (esds) also depend on the quantity of X-ray diffraction data collected (a function of such factors as size of crystal employed, degree of atomic thermal motion in the crystal, and atomic disorder within part of a molecular structure). The overall "goodness" of a crystal structure may be assessed by the crystallographic reliability index, or R-factor. This estimates the agreement between observed and calculated structure factors.

$$R = \frac{\Sigma \left| |F_o| - |F_c| \right|}{\Sigma |F_o|}$$

A typical least-squares highly refined small-molecule structure would have an R-factor of between 0.03 and 0.08. Significantly higher values are indicative of poor crystal quality, small crystal size perhaps, or of errors in the model. The derived esds in, for example, bond length range from about ± 0.002 Å for a highly refined small molecule structure from a good quality crystal with 0.85 Å high-resolution X-ray diffraction data, to about ± 0.04 Å from a poor quality crystal. Such a crystal structure would have a corresponding esd in torsion angle of 2 to 3°. Even a poorly determined crystal structure would have its overall conformation (in terms of features such as ring pucker) determined to a high degree of reliability, in contrast to the more detailed aspects of individual bond lengths and angles.

Comparisons between features in structures determined by X-ray crystallography are made in terms of individual geometric parameters. It is thus important that these are performed in meaningful statistical terms. For example, it is common practice to compare two bond lengths r_1 and r_2, in order to ascertain whether they are significantly different. It has been shown that $|r_1 - r_2|$ is significant at more than the 99% confidence level if it is more than 2.5 times $\sqrt{\delta(r_1)^2 + \delta(r_2)^2}$ (δ = esd) however, since even well-refined X-ray crystal structures typically contain some unaccounted systematic error, this significance factor should be somewhat higher.

The most extensive compilation of X-ray crystallographic data is the computerized data base of some 50,000 organic compounds set up by the Cambridge (U.K.) Crystallographic Data Centre.[36] Programs are available for both accessing the primary data and for extensive statistical analysis of it. As yet, this facility has not been used for the comparative examination of polycyclic hydrocarbon metabolite structures, possibly because of their relatively small number and only moderate accuracy in some instances.

The following sections describe the crystal structures published to date in this area. Their relative paucity can be attributed to two factors; (1) instability of some metabolites and (2) difficulty in obtaining pure metabolites in quantities sufficient for crystallization. Some of these crystal structures are of low accuracy due to poor crystal quality and small crystal size.

B. The Structures

This section reviews metabolites of PAH, their adducts with nucleic acid fragments, and model systems for them. Acetylaminofluorene structures are discussed in Chapter 5 of this volume. Table 1 summarizes these former structures. The structures of the hydrocarbons themselves are not detailed here as their relevance to carcinogenic processes is only indirect by comparison with their activated metabolites.

1. Diols

The first three structures in Table 1 are all non-K-region-hydroxylated. In all cases the

Table 1
X-RAY CRYSTALLOGRAPHIC ANALYSES OF METABOLITES OF PAH

	R factor	Mean esd in bond length	Ref.
Diols and tetrols			
(±)-1,2-Dihydroxy-BA	0.069	0.002	37
(±)-10,11-Dihydroxy-BA	0.059	0.004	37
(±)-7,8-Dihydroxy-BP	0.074	0.005	38
(±)-5,5-Dihydroxy-7,12-DMBA	0.063	0.005	39
(±)-8,9,10,11-Tetrahydroxy-BA	0.069	0.01	40
(±)-7,8,9,10-Tetrahydroxy-BP	0.083	0.01	41
Epoxides and diolepoxides			
(±)-5,6-Epoxide of DMBA	0.043	0.003	42
(±)-9,10-Epoxide of phenanthrene	0.063	0.006	42
(±)-4,5-Epoxide of BP	0.065	0.005	43
(±)-3,4-Epoxide of cyclopenta(cd)pyrene	0.105	0.009	44
(±)-Anti-7,8-dihydroxy-9,10-epoxy-BP	0.106	0.02	45
(±)-Syn-7,8-dihydroxy-9,10-epoxy-BP	0.089	0.013	46
(±)-Anti-3,4-dihydroxy-1,2-epoxy-naphthalene	0.031	0.009	47
Syn-3,4-dimethoxy-1,2-epoxy-naphthalene	0.037	0.003	48
(±)-Syn-2-methyl-3,4-epoxy-1-naphthol	0.091	0.004	49

Abbreviations: BP: benzo(*a*)pyrene; BA: benzo(*a*)anthracene; and DMBA: dimethylbenz(*a*)anthracene.

vicinal hydroxyl groups are *trans* to each other. Whereas they are diaxial in the bay-region 1,2-dihydroxy-BA, they are diequatorial in the nonbay-region 10,11-dihydroxy-BA and 7,8-dihydroxy-BP. Parallel NMR studies in solution have shown that this bay-region conformer is the sole one present, whereas the nonbay-region structures show axial-equatorial equilibria in solution. This difference is primarily due to steric interactions in the bay region; the proximity of the hydrogen atom at the 12 position to the hydroxyl group attached to the C1 atom forces it to be axial with consequent hindrance to ring repucker and axial-equatorial interconversion. A recent *ab initio* quantum chemistry calculation[50] has confirmed that the diequatorial conformation is the most stable one for bay-region vicinal diols of PAH.

A recurrent observation in these structures is the introduction of olefinic bond character adjacent to the hydroxyl groups, compared to the parent hydrocarbons. Thus, further hydroxylation or ring epoxidation could be expected to occur at these electron ring regions. This is indeed the case, as observed for example, in the case of BP: 7,8-diol-BP → 7,8-diol-9,10-epoxy-BP. The C9-C10 distance in the 7,8-diol of BP, of 1.32 Å is 0.036 Å (i.e., 9 esds) shorter than in the parent hydrocarbon. In 1,2-dihydroxy BA,[37] it is the 3,4 bond that is shortened to 1.332 Å. Several of the BP and BA diols (and possibly their other hydroxylated metabolites), show characteristic K-region bond shortening, as does the parent hydrocarbons. It has been established in the case of BP, that the K-region 4,5-epoxide is a primary metabolite and may be produced in detoxification processes.[7] Hydroxylations to produce phenols at various positions also occur during metabolic activations, although few of them have been fully characterized. The crystallographic results indicate that carcinogenic polycyclic hydrocarbons are sensitive to geometric changes produced by partial ring saturation and hydroxylation, with resultant stereoelectronic changes remote from the initial sites of attack. At the present time, the relatively small body of crystallographic data does not enable this issue to be fully analyzed. The crystal structure of the K-region diol of 7,12-dimethyl BA[39] is notable in being of a synthetic *cis* rather than the *trans* series produced enzymatically. Severe K-region ring buckling was observed in this structure, which the authors of the study suggest would have been more extensive if the 6-position hydroxyl group had been equatorial rather than the observed axial conformation.

Table 2
RING TORSION ANGLES, IN DEGREES, IN VARIOUS METABOLITES OF PAH, AS DETERMINED BY X-RAY CRYSTALLOGRAPHY

Structure	α	β	γ	δ	ε	μ	Ref.
BP 7,8-diol	−9.1	−14.9	1.6	33.7	−54.2	44.1	38
BP 7,8,9,10-tetrol	−3.7	14.7	−14.4	65.9	−53.8	22.8	40
Anti BP 7,8-diol 9,10-epoxide	6.6	−19.4	−1.9	34.0	−44.2	24.8	45
Syn Bp 7,8-diol 9,10-epoxide	4.9	−22.6	−2.5	45.3	−59.2	37.0	46
BA 8,9,10,11-tetrol	7.7	15.8	−49.0	61.6	−38.5	4.2	41

The ring torsion angles are designated as shown:

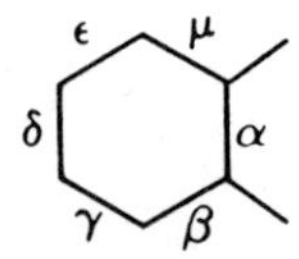

2. Tetrols

The crystal structures have been determined from the hydroxylated ring-opened tetrols that may be obtained directly from hydrolysis of the reactive anti "ultimate carcinogen" diol epoxides of both BP and BA. Their stereochemistries are directly relevant to those when covalently bound to the N2 atom of guanine in DNA. In both cases, the racemic structures have the same relative orientations of four hydroxyl group substituents, and in particular the *trans* orientations of the two hydroxyl groups replacing the epoxides confirm the *trans* ring-opening of this ring. There is evidence from a number of studies that this involves an intermediate carbonium ion. The relative orientations of the hydroxyl groups in the BP tetrol[40] are 7α, 8β, 9β, and 10α, which are equivalent to the 8α, 9β, 10β, and 11α in the BA tetrol.[41] However, the conformational properties of the two structures are dissimilar, with the former having 7(e), 8(e), 9(a), and 10(a) and the latter 8(a), 9(a), 10(e), and 11(e). Solution evidence from proton NMR indicates that the BA tetrol is an approximate 1:1 mixture of two conformers, one of which is the crystallographically observed one, and the other being the e,e,a,a. The crystallographically observed conformation for the BP tetrol would be expected to be preferred in solution, since the axial hydroxyl group at C10 minimizes steric repulsions between it and the hydrogen atom attached to C11. This conformation is in accord with NMR and chemical assignments.

The saturated rings in these polycyclics adopt distorted half-chair puckers. Table 2 and Figure 5 show that quantitatively the puckers in the BP and BA tetrols are very similar, with the major out-of-plane ring atoms being highlighted. In both, atoms (**) deviate more from the least-squares plane of the ring than atom (*).

3. Epoxides

Table 3 details the epoxide ring geometries found in these structures. Average C1–O and C2–O bond lengths are 1.456 and 1.467 Å, respectively. These are significantly longer than values found from a survey of the Cambridge Crystallographic Data base,[51] and may be a reflection of the differences in electronic character in the proximities of the epoxide rings in these diverse structures. Asymmetry of the C–O bond lengths overall is indicated in the table, although the high esds in some structures make individual comparisons of dubious significance.

The consistent trend in the two anti and syn BP structures of bond length C1–O being longer than C2–O, when taken together, is noteworthy, especially since it is in accord with

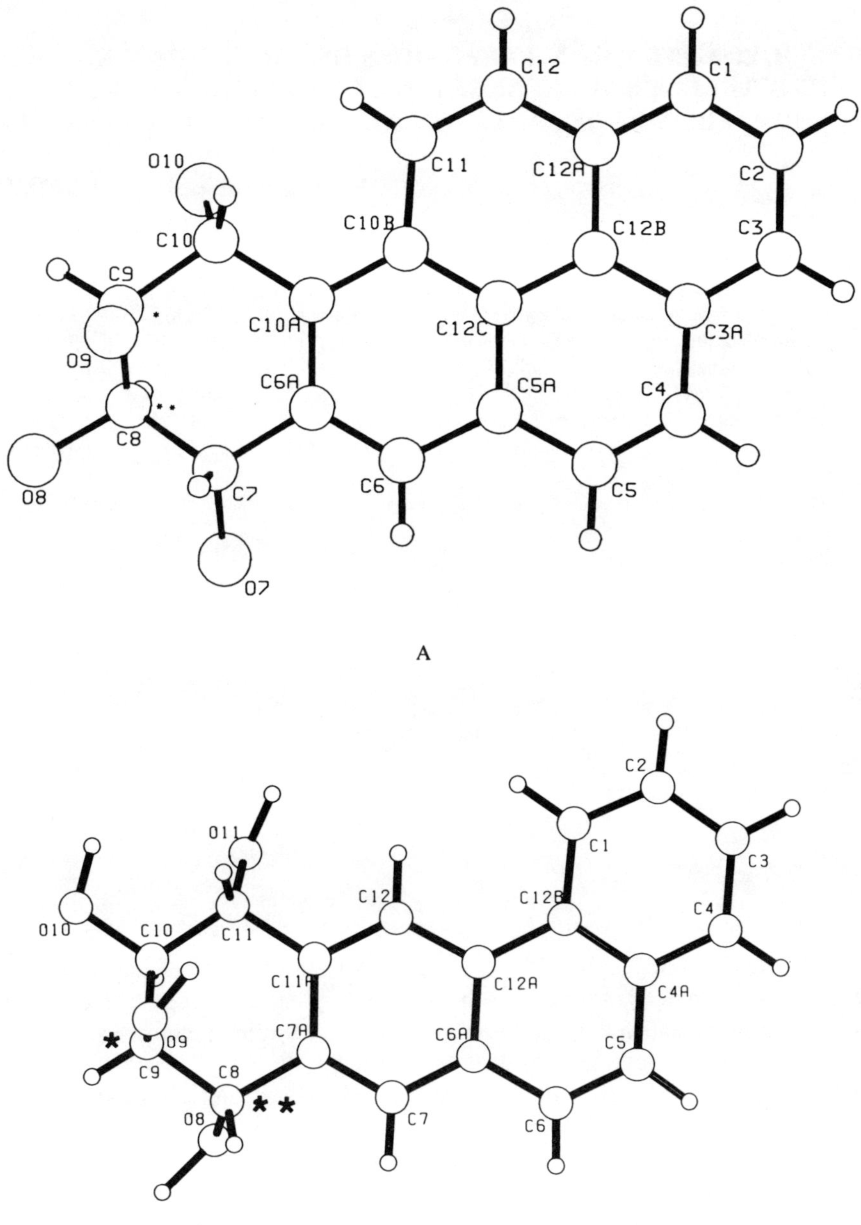

B

FIGURE 5. The crystal structures of (A) BP and (B) BA tetrols. An atom with ** deviates more from the least-squares plane of the saturated ring than an atom with *.

the established epoxide ring-opening pattern in this series. This involves carbonium ion formation on the more electrophilic atom, and the epoxide oxygen atom being retained on C9, in due course as a hydroxyl group (Figure 3). This C–O bond length asymmetry is in the opposite sense to that found in the epoxides of naphthalene,[47-49] which do not possess a bay-region. The notion that bay-region-induced steric forces are responsible for the BP epoxide asymmetry is not in accord with force-field calculations[52] that found both C–O bond lengths to be 1.430 Å in length. It may however be significant that the crucial bay-region

Table 3
GEOMETRY OF THE EPOXIDES OF PAH, FROM X-RAY CRYSTALLOGRAPHIC ANALYSES. BOND LENGTHS ARE IN Å, AND ANGLES IN DEGREES WITH THEIR esds IN PARENTHESIS

	C1–O	C2–O	C1–O–C2
(±)-5-6-Epoxide of DMBA	1.445(3)	1.457(3)	60.4(2)
(±)-9,10-Epoxide of phenanthrene	1.461(6)	1.459(6)	60.6(4)
(±)-4,5-Epoxide of BP	1.478(5)	1.481(5)	60.3(2)
(±)-3,4-Epoxide of cyclopenta(cd)pyrene	1.448(10)	1.496(8)	61.3(4)
(±)-Anti-7,8-dihydroxy-9,10 epoxide of BP	1.460(15)	1.446(15)	59.2(15)
(±)-Syn-7,8-dihydroxy-9,10 epoxide of BP	1.507(12)	1.474(12)	61.0(10)
Anti-3,4-dihydroxy-1,2-epoxy-naphthalene	1.412(8)	1.468(7)	60.6(4)
Syn-3,4-dimethoxy-1,2-epoxy-naphthalene	1.432(3)	1.448(4)	60.9(2)
(±)-Syn-2-methyl-3,4-epoxy-1-naphthol	1.466(4)	1.473(3)	60.0(2)

Atomic labeling is as shown

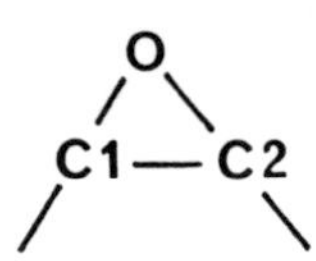

In the cases of epoxides not in the bay region, the arrangement is as shown

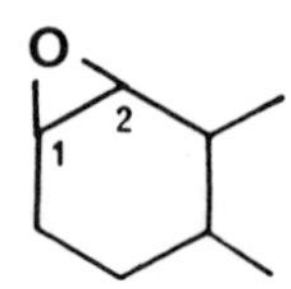

H10 . . . H11 nonbonded hydrogen . . . hydrogen distance was calculated to be 0.46 Å longer than the short 1.69 Å found in the anti diol epoxide,[45] suggestive of considerable extra strain in the actual molecular structure. Earlier quantum chemical calculations using the MINDO/3 approximation,[53] resulted in unusually short C–O epoxide bond lengths; however, this may be an artifact both of the MINDO/3 method, and of the model cyclohexene system used.

The conformational properties of epoxides of PAH, especially of BP have long been a topic of interest and controversy.[19] The anti-7,8-diol-9,10-epoxide[45] crystal structure (Figure 6) shows, in the solid state, that the two hydroxyl groups are *trans* to each other, and are both equatorial (Figures 5 and 6). The epoxide group thus has an axial conformation. This e,e,a arrangement had previously been predicted on the basis of nonbonded contacts.[37,54] It is also consistent with several theoretical studies on the relative stabilities of different conformers which all indicate that the lowest energy conformer corresponds to the crystallographically observed one[53-55] and explains NMR data.[56] The overwhelming evidence of the observed conformation being the lowest energy one, is important for subsequent computer modeling studies on its interactions with DNA; in any case it is computationally not feasible (except perhaps by a molecular dynamics approach) to consider all conformational possibilities in this case.

The (±) syn isomer of BP diolepoxide has high reactivity (see below). It has been suggested[19] that this is due to a conformation that could promote nucleophilic attack on the epoxide ring, and theoretical studies have generally found that the preferred conformation would have an a,a,e conformation (Figure 7) such that an intramolecular hydrogen bond

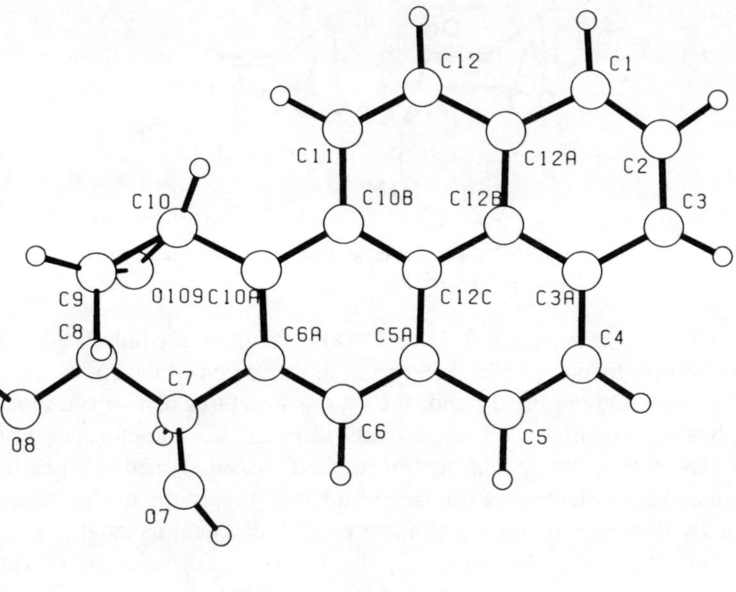

A

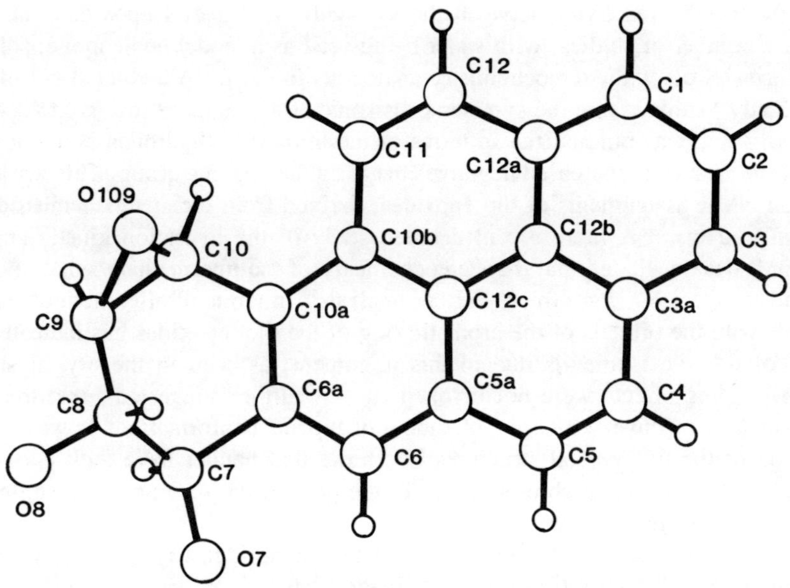

B

FIGURE 6. The crystal structures of (A) anti and (B) syn BP 7,8-diol-9,10-epoxides.

between the hydroxyl group on C7, and the epoxide group, by means of calculations of the relative stabilities of the two syn conformers.[53-55]

The X-ray crystallographic analysis of the (±)-syn BP diol epoxide itself,[46] however, shows the alternative e,e,a conformation (Figure 6). Thus, the proposed intramolecular hydrogen bond cannot be formed. The apparently severe steric interactions between the hydrogen atom on C8, and the epoxide oxygen atom, are in actuality relieved by a combination of cyclohexene ring pucker changes and movement of the epoxide oxygen atom,

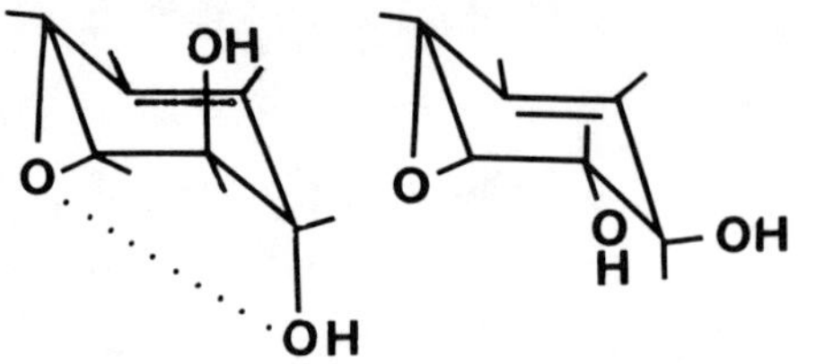

FIGURE 7. Possible conformations of the isomeric BP 7,8-diol-9,10-epoxides.

so that the H8 . . . O distance is 2.55 Å. This is quantitatively illustrated in Table 2, which shows that the ring torsion angles differ most in the region of the epoxide group, compared to the anti isomer. On the other hand, the crystal structures of two syn model compounds, both naphthalene derivatives,[48,49] consistently show the a,a,e conformation, one of which[49] shows the expected intramolecular hydrogen bond. As mentioned in an earlier section, this difference may be a reflection of the lack of a bay region in the model compounds. Further experimentally derived structures and more exact theoretical calculations should be useful in illuminating this issue. In particular, the known deficiencies of several approximate quantum chemistry procedures need to be overcome, preferably by reliable *ab initio* methods. The increasing availability of supercomputers should make this task a realizable one.

The difference in reactivity between the syn and anti BP diol epoxides, has been the subject of a number of studies, with water being used as a model nucleophile; solvolysis at C10 produces tetrols with stereochemistry analogous to the DNA adduct at N2 of guanine. An early study[57] showed that the syn isomer is some 30 times more reactive to spontaneous hydrolysis, which was rationalized in terms of intramolecular hydrogen bonding providing anchimeric assistance to increased negative charge on the epoxide group. This was supported by solution NMR assignments of the epoxides, derived from the stereochemistries of their hydrolysis products.[29] A more recent detailed study of the hydration kinetic and product distribution[58] has concluded that the stereochemistry of the intermediate syn carbonium ion corresponds to the e,e,a conformer. Further analysis[59] in terms of alignment of the epoxide C–O bonds with the orbitals of the aromatic ring of the diol epoxides has indicated that the reactivity of the syn isomer is due to this alignment, as seen in the crystal structure.[46] Hydrogen bonding effects were not invoked. It was further suggested[59] that the anti conformers are less reactive by virtue of their nonaligned conformation; however, detailed examination of the two crystal structures[45,46] shows that neither have their epoxide group orthogonal to their aromatic chromophore. In the case of the (±) anti, the angle involved is 83° and 118° for the (±) syn.

An illustration of the potential usefulness of the Cambridge Crystallographic Data base for studying intermolecular interactions of epoxides, has been recently given.[60] Comparative examination of hydrogen bonding and other interactions in all relevant crystal structures has shown that this occurs in a plane perpendicular to the epoxide group, with a relatively broad distribution of hydrogen bond position. The relevance of this approach to considerations of stereochemistry and interaction with DNA has not as yet been explored, although some modeling studies (see below) have implicitly used this promising approach.

III. DNA — CARCINOGEN ADDUCTS

To date no X-ray crystallographic structural studies have been reported on oligonucleotides alkylated by carcinogenic compounds, such as BP 7,8-diol-9,10-epoxide. Eight structures of alkylated nucleosides have been reported.[61,62] However, the alkylation site of these compounds is either at N6 of adenine or O6 of guanine, and as we will see later, they are not directly related to the active adducts. Computerized molecular modeling, on the other hand, provides a powerful tool to study in principle all categories of interactions between DNA and PAH. Several molecular modeling studies have been carried out on BPDE-DNA adducts using short (up to tetramer) oligonucleotide duplexes in order to explain the biophysical and biochemical experimental data. The background to these models and the underlying experimental data are described in the following section.

A. Experimental Background

1. Sites of Alkylation by BPDE

Electrophilic mutagens and carcinogens generally form covalent bonds to nucleophilic site on DNA. The N7 site of guanine possesses relatively high nucleophilic strength compared to hydroxyl groups and amino groups. Electrophiles reacting through an SN2 mechanism (such as dimethylsulfate) usually prefer higher nucleophilic sites such as N7 of guanine; by contrast, compounds reacting mainly via an SN1 mechanism (e.g., *N*-methyl-*N*-nitrosamine) are usually less selective for nucleophilic strength and may cause alkylation at hydroxyl groups.[63-66] According to Moschel et al.[67,68] the greater the positive charge density at the reaction site, the more extensive is the reaction at an exocyclic oxygen as compared to exocyclic nitrogen atoms. However, the application of these simple rules has little significance to the reaction of more complicated compounds such as PAH. DNA adducts formed with BPDE in vivo and in vitro have been characterized by a number of workers.[69-75] It has been found by means of high pressure liquid chromatography (HPLC), that the extent of binding of (+)-anti BPDE is approximately 90% at N2 of guanine, 5% at N6 of adenine, 1% at N4 of cytosine, with 4% unidentified. There seems to be some reactions with the N7 position of guanine;[76,77] however, the N7 adduct is labile and difficult to detect from in vivo samples. O6 of guanine and N6 of adenine derivatives were thought[75] to arise mainly from the (−)-anti BPDE (59% at N2 of guanine, 21% of guanine, 18% to N6 of adenine). However, the relative binding to DNA of the (+)- and (−)- isomers is 10:1.[78] This stereoselectivity arises at least in part from the steric factors which generally dominate interactions between ligands and DNA. As is shown in Figure 8, the main alkylation site, N2 of guanine is in the minor groove, while N6 of adenine is in the major groove of standard DNA conformations. Interestingly, the level of binding to N2 of guanine is correlated to the mutagenic activity in mammalian cells, when the (+)- and (−)-anti isomers are compared.[20]

2. Biophysical Experiments

Structures and conformations of adducts between BPDE and DNA have been extensively investigated using a variety of biophysical techniques. These include absorption, linear dichroism, circular dichroism (CD) spectroscopy, fluorescence quenching, and mass spectroscopy (MS). There are two general classes of binding models for the adducts. In the first, the PAH lies in one of the grooves of DNA. This is termed "external binding". The second model, where the chromophore of the PAH is intercalated between the DNA base pairs within the helix of DNA, is called "internal binding". Some of the biophysical data are, at least apparently contradictory. For example, Lefkowitz et al.[80] performed an optically detected magnetic resonance study which showed that the (±)-anti BPDE covalent adduct covalently bound to DNA lies outside the helix. Prusik and co-workers[81,82] find with fluorescence measurements that the BPDE is bound externally in the minor groove and produces only a minor distortion of B-DNA geometry. On the other hand, Drinkwater et al.[83] reported data supporting a model where the carcinogen intercalates between DNA base pairs, forming a stable double-stranded complex.

Two different types of binding mode have been detected by linear dichroism and absorption spectra.[84-86] The first site is characterized by a relatively large red shift of 10 nm in the absorption spectrum around 337 and 354 nm. The corresponding linear dichroism peak is

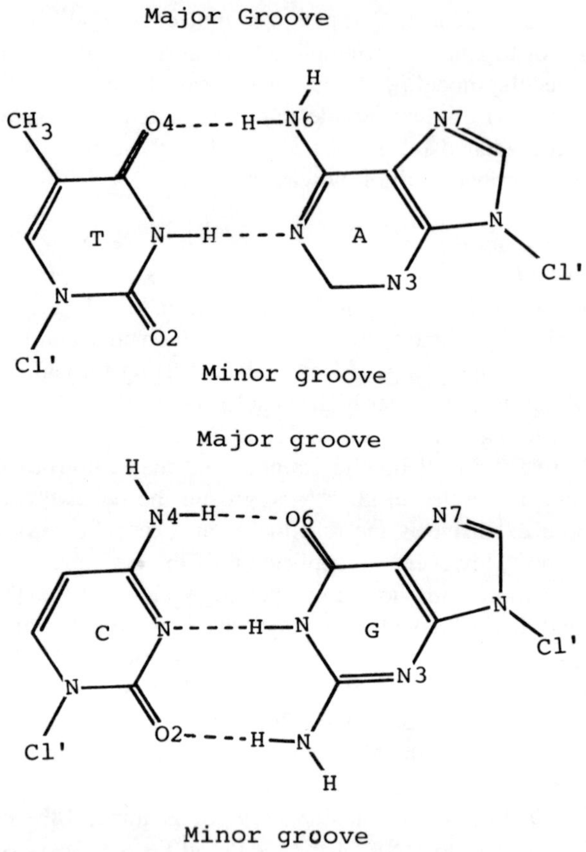

FIGURE 8. Minor and major groove sides of base pairs in the G–C and A–T Watson-Crick hydrogen bonding.

negative. These findings suggest[84] an intercalation geometry where the planar pyrene moiety of BPDE is almost parallel to the planes of the DNA base pairs. The second site is characterized[85] by a much smaller red shift (2 to 3 nm) in the absorption spectrum and a positive linear dichroism at 346 nm. Here the long axis of the pyrene ring-system is tilted up to 35° to the DNA helix.[87] These two classes of binding, termed by Geacintov et al.,[85] as Site I and Site II, correspond, respectively, to the internal and external binding mentioned earlier.

Work by Undeman et al.[88] has clearly shown that the structure of the adduct formed between DNA and BPDE depends on the stereoisomers of the hydrocarbons. Futhermore, the reaction seems to proceed in more than one step. Utilizing kinetic methods,[89,90] it has been shown that formation of an internal complex occurs within 5 msec or less when two aqueous solutions of BPDE and DNA are mixed together, whereas the BPDE alkylates DNA on a time scale of minutes. For the (±)-anti BPDE, the negative linear dichroism spectrum at an early stage of reaction eventually converts to a positive peak. However, for the (±)-syn BPDE, linear dichroism ($A_{\parallel}$-$A_{\perp}$) continues to be negative.[88,90] The disappearance of intercalated (+)-anti BPDE adducts was reported[91] to be directly proportional to the rate of appearance of covalent adducts. These results suggest that reorientation of the pyrene moiety occurs after an initial intercalative binding, in the case of the (±)-anti isomer. On the other hand, conformational change appears to be minor for the (±)-syn BPDE complex. The two enantiomers of anti-BPDE behave differently. A detailed analysis of linear dichroism using (+)- and (−)-anti BPDE separately shows[92] that the (+)-enantiomer predominantly forms an externally bound adduct (90%) with a 10% contribution from internal binding. The (−)-

isomer adducts on the other hand, seemed to be more equal mixtures of both internal and external types. The orientation angle, α(BPDE), which is defined as an angle between the long axis of the pyrene moiety and the average DNA helix axis is in the range of 15 to 30° for the external complex of the (+)-anti isomer. In the case of the (−)-adduct, the pyrene moiety of external binding seems to be more tilted, with a value of 37 to 45°.[92] The orientation angle of the internal binding complexes in the (−)-adducts lie in the range of 61 to 79°. The angle is much larger for the classical intercalation geometry.

3. Biochemical Studies

Unwinding of closed circular DNA and enzymatic digestion of DNA alkylated by carcinogens provide insight into DNA-carcinogen interactions. Unwinding angles of closed circular DNA by BPDE have been studied by various workers and the values reported are widely spread. Gamper et al.[93] determined the unwinding angle associated with BPDE alkylation, by using superhelical and partially relaxed SV40 DNA. The unwinding angle for superhelical DNA ranged from 33 to 60° depending on the carcinogen/DNA molar ratio. This extremely large unwinding angle must reflect localized denaturation at the site of alkylation. The unwinding angle for the partially relaxed DNA is 30° irrespective of the molar ratio. This suggests that the modified base pairs are disrupted and in the presence of torsional strain act as a nucleus for further denaturation of up to eight adjacent base pairs. However, it was reported[94] that binding to adenine causes the local denaturation of DNA, whereas the guanine adduct which has more than ten times its binding constant did not create such local denaturation. The effect of strand scission by BPDE must be taken into account in the interpretation of unwinding experiment: the reaction at N7 of guanine is known to lead to ring opening and strand scission.[95] Agarwal et al.[96] have shown using relaxed circular PBR322 DNA, that there is a rapid process which increases the mobility of treated DNA and a slower process of strand breakage. By employing a short incubation time with BPDE and rapid electrophoretic analysis, Drinkwater et al. could minimize strand scission and shows that BPDE unwinds and then, at the higher concentration of the carcinogen, overwinds closed circular superhelix SV40 DNA. Under their experimental conditions denaturation of DNA appears not to be the major mechanism for the unwinding of superhelical DNA. The unwinding angle was determined to be 22 ± 3° per bound residue. This is close to the unwinding angle of 26 ± 2° for the noncovalent intercalator ethidium bromide. The changes in superhelical density of the SV40 treated with BPDE were also confirmed by electron micrographs. The relatively low sensitivity of BPDE-DNA adduct to S1 nuclease digestion or formaldehyde modification[94] circular dichroism studies of BPDE-modified GpU[97] are consistent with an internal intercalation mode. Contrary to this, Pulkrabek et al.[98] observed that the binding of BPDE to DNA enhances the susceptibility of the helix to denaturation by heat or formaldehyde and produces an enhanced susceptibility of the modified DNA to S1 nuclease digestion. These results favor an external binding model. The unwinding angle determined by Meehan et al. is 13° for the (+)-anti BPDE adducts.[89] An attempt has been made to resolve these rather contradictory experimental results using electric dichroism, circular dichroism, fluorescence quenching, ethidium bromide measurements, thermal denaturation, and S1 nuclease digestions.[99] It has been observed that at $\gamma = 0.04$ (γ = adducts per base pair), individual DNA segments in the modified DNA are tilted by 29° relative to the axis of orientation of unmodified DNA. No evidence was found for local denaturation of the DNA. Accordingly, a "kink" model has been proposed[99] where the covalently bound BPDE intercalates between DNA base pairs in a wedge-shaped form.

B. Computer Modeling of BPDE-DNA Adducts

As various biophysical and biochemical experimental results and their interpretation suggest two general types of binding mode, (i.e., internal and external binding), molecular modeling has been carried out in both categories. Most studies have employed the X-ray studies outlined above, for definitions of the polycyclic aromatic part of bound complexes.

1. External Binding

Hingerty and Broyde[100] have carried out semiempirical potential energy calculations to investigate the (+)-anti adduct with dCpG. Partial charges for the BPDE and the linking portion of guanine were computed by the INDO method. Varying ten conformational angles (the two glycosyl or glycosidic angles, six dinucleoside backbone torsional angles, and the two linkage angles about the C2–N2 bond of guanine together with the N2 (guanine)-C10 (BPDE) bond), about 1000 trial conformations were examined. The lowest energy conformation obtained had stacking between the hydrocarbon and the adjacent cytidine. This orientation can occur in the B-DNA conformation family with the alteration of guanine glycosidic angle to high anti value of 135°. In standard B-DNA the angle is in the range of 60 to 80°. This partial geometry may produce internal binding (discussed below). The next two lowest energy conformers have the guanine and the cytidine approximately stacked on top of each other, as in normal B-DNA. The long axis of the pyrene is either about 25 to 30° tilted from, or perpendicular to the DNA helix axis. The former conformation may place the pyrene moiety in the minor groove of DNA. In this study, only a single-stranded dinucleoside monophosphate was considered, as this approach requires an extremely large amount of computation time for even moderately sized systems. However, it is important to ascertain whether the hydrogen bonding between the DNA base pairs is conserved or not, and to examine how the adduct dinucleoside unit relates to a longer sequence of duplex DNA. It is interesting to note that in these calculations,[100] Z-type conformation with syn-guanine is not found in the lowest 5 kcal/mol^{-1} level of solutions. This is in contrast to qualitative modeling,[101] which assumed a Z-DNA conformation with syn-guanine, based on several X-ray crystal structure analyses of PAH-modified nucleosides.[61,62] The compounds are the adducts formed between N6 of adenosine or deoxyadenosine and chloromethylanthracene, methylanthracene and methylbenzo(*a*)anthracene, and adducts of guanosine with *para*-substituted benzyl derivatives alkylated at the O6 position. In the crystal structures of all the compounds, the glycosidic angle of the alkylated purine bases was found to be in the syn range. Hence, chloromethyl BA was locked onto a guanine base of duplex d(C<u>G</u>C) (underline indicates the alkalated base) by superimposing the N6 of adenine onto the O6 of guanine. This is a major groove alkylation. A further rotation of 180° about the C6–N6 bond of adenine places the aromatic hydrocarbon on the surface of this DNA helix. The normal interbase hydrogen bonds within the helix are conserved. The large anthracene ring system fits in the major groove. The slight curvature of the dimethylbenz(*a*)anthracene conforms to some extent to the outer shape of the DNA helix. This adduct is a minor one compared with the N2 guanine one. In addition, the hydrocarbon examined is of 7-chloromethyl 12-methylbenz(*a*)anthracene linked through the 7-methyl group and not of the highly mutagenic 7,12-dimethylbenz(*a*)anthracene (DMBA). DMBA is known to be activated *via* metabolites, including its K-region 5,6 epoxides and DMBA-3,4-diyhydrodiols. Interactions of DMBA-5,6-oxide with DNA have been studied by NMR, mass spectra, and circular dichroism. It has been shown[102,103] that the covalent adducts consist of two diastereomeric *trans* adducts between N2 of guanine and the 5- and 6-positions of DMBA. Molecular modeling of this type of adduct may be more relevant to the mutagencity caused by polycyclic hydrocarbons.

External binding of BPDE has been modeled by a number of investigators. Taylor et al.[104] assumed an anti → syn transition of the affected guanine residue. The DNA molecule is flexible and also constantly denatures and renatures. This can be seen in the case of classical intercalation of fully dis-substituted anthraquinones[105] and 1,8-naphthlylimide.[106] These compounds with two bulky substituents on each end of a flat aromatic molecule cannot be inserted directly into an intercalation site. The observed intercalation can only be explained by a local denaturation of DNA prior to the intercalation event. During the course of local denaturation DNA guanine bases can in principle rotate about their glycosidic bonds. Unlike the modeling mentioned earlier,[101] where a Z-DNA conformation was adopted, these studies have examined the possibility of regaining a B-DNA conformation.[104,107] A reasonable low-energy geometry was obtained with a duplex tetramer d(GC<u>G</u>C), adopting a B-DNA conformation except for the glycosidic angle of the next guanine residue. By rotating the angle by 200° from the normal anti to a syn structure, the steric contacts becomes minimal and the pyrene moiety is placed in the DNA major groove.[81] Hydrogen bonds between the affected guanine and the cystosine in the opposite strand appear to be perturbed. In the modeling into a Z-DNA,[101] the hydrogen bonding was not found to be distorted. Anti → syn rotation of an adenine base displaces the N6 atom only slightly farther into the major groove. Thus, this rotation would not be particularly favored.

An early modeling study using computer graphics employed a standard B-DNA,[108] $dG_3 \cdot dC_3$, using fiber diffraction X-ray coordinates.[109] BPDE was covalently bound to the N2 of a central guanine residue and only the two rotations about the C10(BPDE)-N2(guanine) bond and the N2–C2 bond of the guanine were allowed. The contour diagram of close contact indicated that the hydrocarbon can be placed in the minor groove of DNA; however, steric interactions were severe. A similar orientation was found for the low-energy state in the case of the 2-acetylaminofluorene adduct; steric interactions were found to be much less for this less bulky carcinogen. The results suggest that conformational changes in B-DNA itself are required if it is to accommodate a bulky hydrocarbon molecule in the minor groove.

Similar conclusions of severe steric interactions in the minor groove of B-DNA were reached by Aggarwal et al.[110] They adopted a standard A- or B-DNA duplex d(C<u>G</u>C) trimer as a receptor using the atomic coordinates developed from fiber diffraction.[109] A rotation search was first carried out to find regions of low-energy conformation, rotating about the two linkage bonds (C2–N2 and N2–C10). Then, starting at the low energy points, full energy minimization was carried out employing up to 20 torsion angle variables in the 5′-end of dinucleotides of each trinucleotide strand. Full minimization did not change torsion angles by more than 2 to 3° from the starting values. In the optimum geometry, the pyrene ring system is snugly placed in the minor groove of A-DNA, with the orientation angle, α(BPDE), of 67 and 63° for the (+) and (−)-anti isomers, respectively. Figure 9 illustrates the low energy conformation of (+)-anti BPDE in the minor groove of A-DNA. In the case of the (+)-isomer, there appears to be an additional hydrogen bond-like contact between the hydrogen atom attached to C1′– of the 3′-end cytidine and O9 of BPDE. The inability of BPDE to form a low-energy structure with a standard B-DNA is understandable when the dimensions of the minor groove are considered. In B-DNA, the minor groove is only 5.7 Å wide and over 7 Å deep, whereas in A-DNA, it is 11 Å wide and less than 3 Å deep. The BPDE bound to the N2 of guanine has a molecular thickness of about 6.2 Å, so it could not easily reach the relatively buried N2 of guanine in B-DNA.

Hence BPDE adduct formation must be accompanied by unwinding of DNA, or a local B- to A-DNA transition, or an anti to syn transition of the alkylated guanine to place the pyrene moiety in the major groove of DNA. An alternative way to alleviate the severe steric interactions caused by the bulky pyrene moiety in the B-DNA minor groove is to introduce a kink in DNA while retaining the chromophore in the minor groove. Modeling into a kinked DNA has been extensively studied by means of molecular mechanics calculations.[107,111] A kink was introduced at an intercalation site to create an opening in either the major or the minor groove. Starting from a parallel orientation, base pairs above and below the intercalator were rotated by $+\frac{\alpha_x}{2}$ and $-\frac{\alpha_x}{2}$, respectively, about a kink axis which is parallel to the interstrand C1′(py) − C1′(pu) axis. The definition of the kinking parameter α_x, is given in Figure 10. Two types of sugar puckering at the intercalation site, C2′-*endo*-(3′-5′)-C2′-

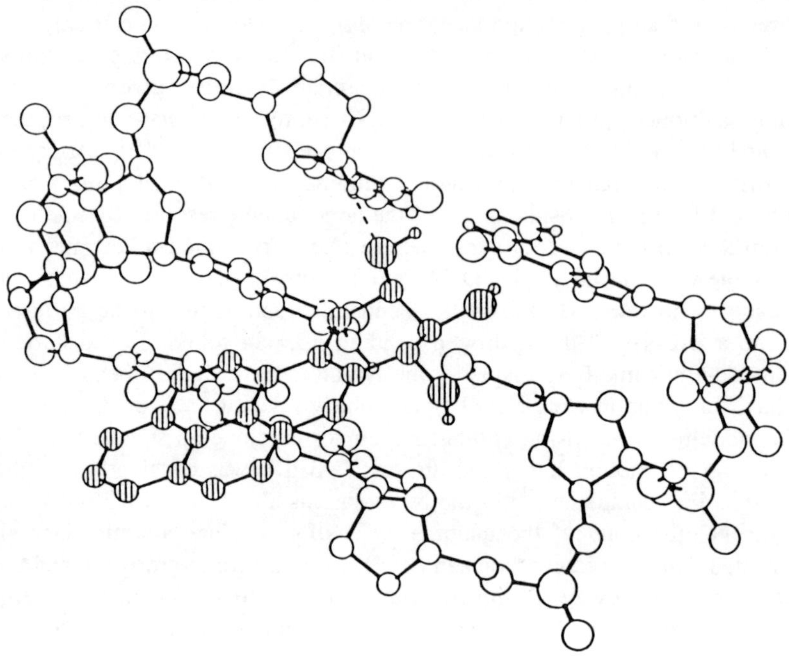

FIGURE 9. (+)-Anti BPDE adduct with A-DNA, the C1′..09 interaction is shown by a dashed line. (From Aggarwal, A. K. et al., *J. Biomolec. Struct. Dyn.*, 873, 1983. With permission.)

endo and C3′-*endo*-(3′-5′)-C2′-*endo*, were examined. An optimum structure was obtained in a kinked duplex d(GCGC) alkylated at the N2 of guanine. The kinked DNA adopts a conformation of $\alpha_x = -70°$ (minor groove opening). The orientation angle of BPDE, α(BPDE), becomes 15° in this structure. The geometry is 10.1 kcal/mol more stable than an intercalative covalent form in the same kinked DNA receptor site (this is discussed below). Likewise, external binding is more favored than internal for the (+)-syn-BPDE-N2 (guanine) adduct, when bound in the kinked DNA. This mode of external binding in a kinked DNA appears to explain the orientation parameters obtained from linear dichroism experiments.

2. Internal Binding

a. Noncovalent Intercalation

As we have seen, some experimental data have suggested external binding of BPDE, that is on the exterior of the double helix. However, other evidence suggests that unwinding is concomitant with binding and that therefore an intercalative binding model is appropriate. Parallel separation of adjacent base pairs permits the BPDE to form a classical intercalation complex[112] which has been found in a number of X-ray crystallographic studies of short oligonucleotide sequences with drugs such as proflavine and ethidium bromide.[54-58] Adjacent base pairs retain a parallel orientation to each other and the chromophore of an intercalator is inserted between the base pairs. This type of classical intercalation complex has been modeled for (+) and (−)-BPDE in two independent studies.[107,118] In the latter,[118] the molecules of (+) or (−)-anti BPDE were simulated to approach the already opened-up site of the self-complementary dinucleoside duplex dCpG by interactive computer graphics modeling, following which, empirical energy calculations were carried out. The receptor geometry was fixed during the docking process and only rotational and translational movements were allowed for the BPDE molecule. For both enantiomers, approach from the major or the minor groove directions resulted in well-defined low-energy intercalative geometries;

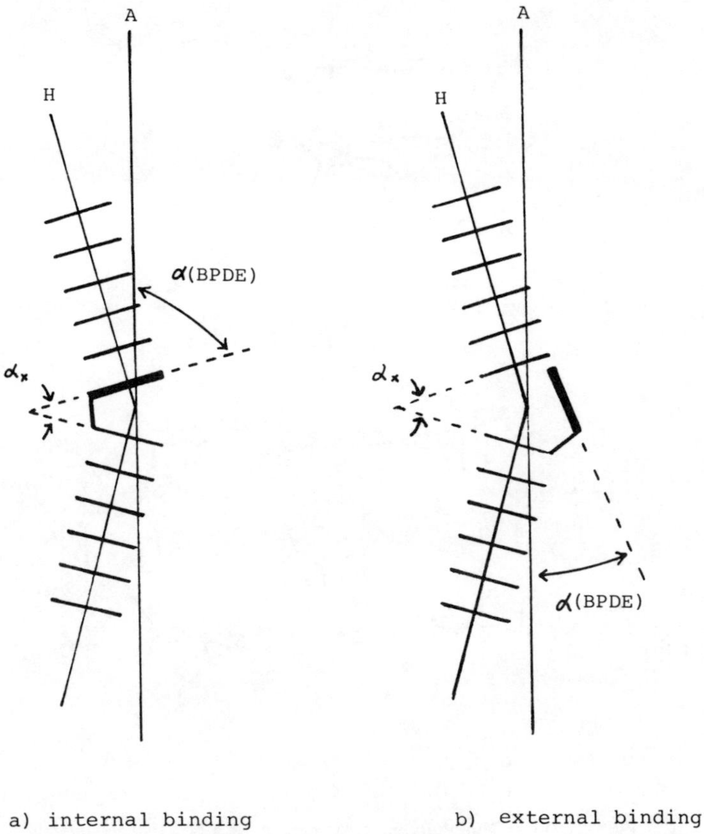

FIGURE 10. Definition of kink angle α_x and the orientation angle α(BPDE). H and A indicate local and average helix axis of DNA. Thick bars represent the pyrene moiety and thin bars represent base pairs.

these are shown in Figure 11. In all cases, the long-axis of BPDE is oriented at right angle to the direction of the Watson-Crick base pairs, as in the case of the drug-hexamer complex, daunomycin-d(CGTACG).[119] The approach from the minor groove brings the N2 of guanine close to the (+)-anti BPDE epoxide oxygen atom (at a distance of 2.82 Å), while the major groove approach brings the N4 of the cytosine to 2.80 Å from the epoxide oxygen. There are similar contacts in the case of the (−)-isomer. Modeling of noncovalent intercalation by Miller et al.[107] used a duplex tetramer d(GCGC) as a DNA model. The intercalation site was dCpG, as in the study detailed above. Three sites with different unwinding angles, 7 to 12°, 14 to 18°, and 25 to 32° were tried, of which the smallest unwinding site was most favored. The optimum intercalation geometry is similar to that reported by Subbiah et al.[118] Thus, minor groove intercalation brings the epoxide close to the N2 of guanine, while for the major groove intercalation, the epoxide oxygen is located near the N4 of cytosine or O6 of the guanine residue. The distinct close interaction between the N2 of guanine and the epoxide oxygen atom found in the modeling is particularly noteworthy, since the covalent bond to the N2 of guanine results in the DNA lesions that have been most frequently correlated with carcinogenicity. Intercalative interaction appears to occur at an early stage of reaction between BPDE and DNA.[89,90] Thus, it is tempting to speculate that these minimum-energy classical intercalation geometries represent a stage prior to either the covalent binding, or subsequent hydrolysis to tetrols. However, the minor groove intercalaton model of (+)-anti-BPDE indicates *cis*-addition, contrary to the established *trans*-geometry of covalent

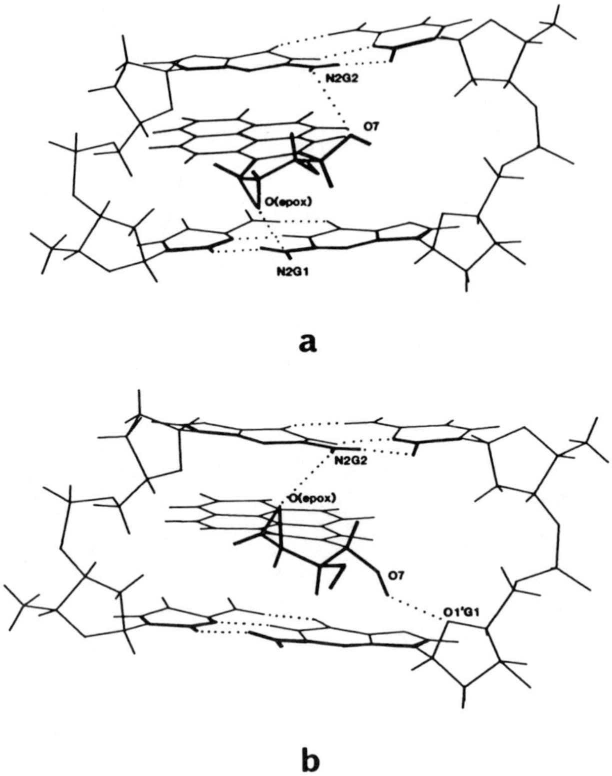

FIGURE 11. Minimum-energy structures of a noncovalent intercalative adduct of (+)-anti BPDE (a) and (−)-anti BPDE (b) with DNA. The view in both cases is into the minor groove of the DNA. (From Subbiah, A., Islam, S. A., and Neidle, S., *Carcinogenesis*, 4, 211, 1983. With permission.)

adducts. Moreover, the distance between the epoxide oxygen and the N2 of guanine alone cannot be a criterion of preferred reaction. In order to form a single bond between the C10 of BPDE and the N2 of guanine, the C10 and N2 atoms must approach, close and adopt sp^3 and sp^2 hybridizations, respectively. This bond formation and the classical intercalation cannot occur at the same time.

b. Covalent Intercalation

Based on the experimental results of unwinding of superhelical SV40 by BPDE and electron microscopy, Drinkwater et al. carried out model-building studies with CPK space-filling models.[83] The results indicate, that after covalent binding of BPDE at the N2 of guanine, an alteration in the pattern of sugar puckering at the site of intercalation, from the normal C2′-*endo* of B-DNA to C3′-*endo*-C2′-*endo*, as in a model for drug intercalation,[116] together with a rotation of the modified guanine residue about the glycosidic bond can keep the pyrene moiety in the interior of the helix. These alterations result in the extension of the sugar-phosphate backbone and in the unwinding of the DNA helix by 20 to 30°. This intercalation model incorporates favorable geometry for stacking interactions between the planar hydrocarbon moiety and the cytosine residue that was formerly hydrogen bonded to the substituted guanine residue and between the pyrene ring-system and the adjacent base pairs. However, it is difficult to assess, from the purely qualitative photograph published of this CPK model, how realistic is the possibility of covalent intercalative binding with the plane of the hy-

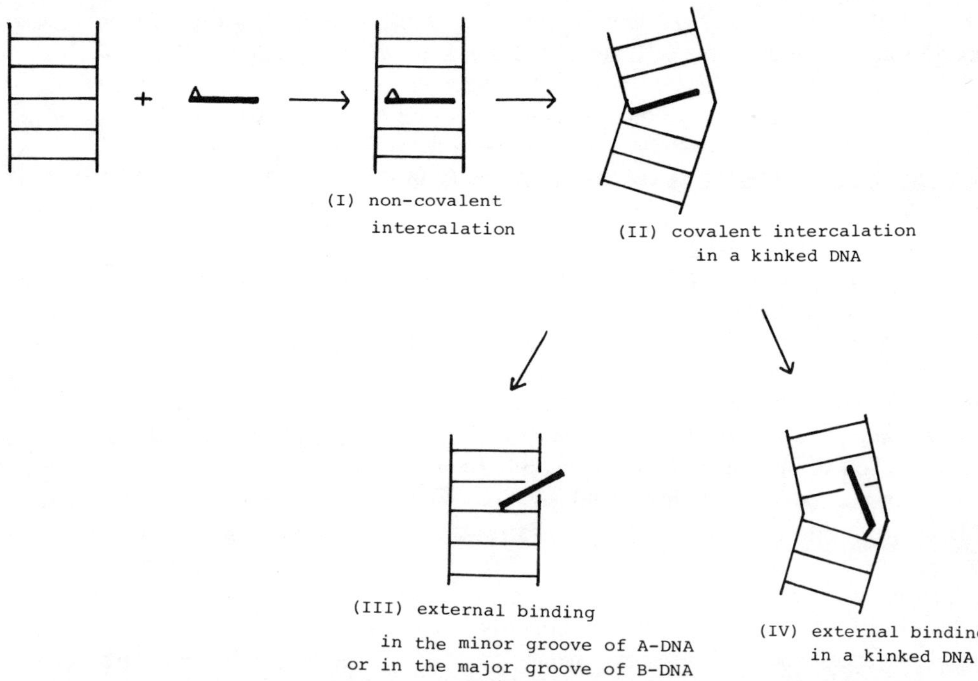

FIGURE 12. Various conformations of BPDE-DNA adducts and their formation pathways.

drocarbon almost perpendicular to the helix axis of DNA. Strictly classical intercalation with concomitant covalent binding at the N2 of guanine is not sterically possible. It becomes feasible only when a kink is introduced at the affected site of DNA. Hogan et al.[99] employed a CPK modeling approach and presented two models which they proposed to explain their own and other workers experimental findings, such as the lack of denaturation of DNA and the protection of the bound carcinogen from interactions with the solvent. If BPDE binds on the 3′-side of a guanine, the pyrene moiety is positioned over a cytosine residue. If BPDE binds to a guanine on the 5′ side, the pyrene moiety becomes positioned over the guanine. This CPK model shows the plane of the BPDE tilted by an angle of 25 to 30° with respect to the plane of the guanine ring in the 3′ complex, and by an angle of about 50° in the 5′ complex.

Molecular mechanics calculations of internal covalent bonding in a kinked DNA have been carried out.[107] The kinked DNA receptor sites were those used for subsequent external binding analysis. Binding to N2 of guanine in a duplex tetranucleotide d(GC<u>G</u>C) kinked in the major groove ($\alpha_x = 39°$) shows that (+)-anti and (+)-syn-BPDE fit in DNA sites better than their enantiomers. On the other hand, the adduct to N6 of adenine in a kinked d(C<u>A</u>TG) model ($\alpha_x = -30°$, minor groove opening) favors the (−) over the (+) enantiomer for both anti and syn isomers. The difference in the energies required to fit each isomer of BPDE in the kinked receptor site is rather large. Therefore, it is proposed that during this intercalative covalent binding process, stereoselectivity in the adduct formation occurs. In the intercalative covalent binding process, the orientation angle, α(BPDE), is larger than 80°. However, this value is significantly greater than observed experimentally. This stage in binding can further relax to produce an external binding mode, the conformation of which appears to correspond more closely to experiment.

C. Summary

These possible intermediates in the reaction of BPDE with DNA are summarized in Figure 12. First, classical noncovalent intercalation occurs (I). This step places the epoxide group in one of the grooves of DNA, thereby arranging an ideal geometry for subsequent covalent bond formation between the C10 atom of BPDE and the appropriate base atoms such as N2 of guanine or N6 of adenine. Then DNA undergoes a local conformational change to accommodate the covalently bound pyrene moiety inside the DNA helix (II). Further rearrangement of the adduct structure occurs to bring the pyrene moiety into one of the grooves of DNA. External binding is possible in A-DNA, in B- (or possibly) Z-DNA (III). In the A-DNA form, the pyrene sits in the minor groove, whereas in the standard B-DNA form, the alkylated guanine changes to, for example, a syn conformation, to place the pyrene ring-system in the major groove. Alternatively, a kinked DNA can place the bulky pyrene moiety in the opened grooves of DNA (IV). Localized conformational relaxation of B-DNA could also conceivably produce a localized groove opening sufficient to accommodate a BPDE molecule. Stereoselectivity of the adduct formation may occur during the intercalative covalent binding step. These models and energetics of adducts and their reaction pathways may explain at least part of the experimental data. X-ray crystal structures of adducts, for example, between BPDE and oligonucleotide alkylates at the N2 of guanine residue are now needed to provide more definitive information; modeling studies on molecular systems of this complexity are not at present capable of unambiguous location of *global* minimum energy structures, as opposed to the *low* energy ones found in the studies reviewed here. The aromatic hydrocarbon-DNA modeling studies reported to date have always incorporated at least some structural information, implicitly in the case of the DNA itself, and explicitly in most studies, in the case of the metabolites. These studies can be considered as first-order approaches to the problems.

REFERENCES

1. **Kennaway, E. L. and Hieger, I.,** Carcinogenic substances and their fluorescence spectra, *Br. Med. J.*, 2, 1044, 1930.
2. **Cook, J. W., Hewett, C. L., and Hieger, I.,** The isolation of a cancer-producing hydrocarbon from coal tar, *J. Chem. Soc.*, 395, 1933.
3. **Phillips, D. H.,** Fifty years of benzo(a)pyrene, *Nature (London)*, 303, 468, 1983.
4. **Hiatt, H. H., Watson, J. D., and Winsten, J. A., Eds.**, *Origins of Human Cancer,* Vol. 1 to 3, Cold Spring Harbor Laboratory, Cold Spring Harbor, N.Y., 1977.
5. **Grover, P. L., Ed.**, *Chemical Carcinogens and DNA,* Vol. 1 and 2, CRC Press, Boca Raton, Fla., 1979.
6. **Gelboin, H. V. and Ts'o, P. O. P., Eds.**, *Polycyclic Hydrocarbons and Cancer,* Vol. 1 and 2, Academic Press, New York, 1978.
7. **Searle, C. E., Ed.**, *Chemical Carcinogens,* Vol. 1 and 2, American Chemical Society, Washington, D.C., 1984.
8. **Singer, B. and Grunberger, D.,** *Molecular Biology of Mutagens and Carcinogens,* Plenum Press, New York, 1983.
9. **Harvey, R. G., Ed.**, *Polycyclic Hydrocarbons and Carcinogenesis,* American Chemical Society, Washington, D.C., 1985.
10. **Pullman, B. and Pullman, A.,** *Quantum Biochemistry,* Interscience, New York, 1963, 155.
11. **Brookes, P. and Lawley, P. D.,** Evidence for the binding of polynuclear aromatic hydrocarbons to the nucleic acids of mouse skin: relation between carcinogenic power of hydrocarbons and their binding to DNA, *Nature (London),* 202, 781, 1964.
12. **Harvey, R. G.,** Activated metabolites of carcinogenic hydrocarbons, *Acc. Chem. Res.*, 14, 218, 1981.
13. **Sims, P., Grover, P. L., Swaisland, A., Pal, K., and Hewer, A.,** Metabolic activation of benzo(a)pyrene proceeds by a diol-epoxide, *Nature (London),* 252, 326, 1974.
14. **Sims, P. and Grover, P. L.,** Epoxides in polycyclic aromatic hydrocarbon metabolism and carcinogenesis, *Adv. Cancer Res.*, 20, 165, 1974.
15. **Newbold, R. F. and Brookes, P.,** Exceptional mutagenicity of a benzo(a)pyrene diol epoxide in cultured mammalian cells, *Nature (London),* 261, 52, 1976.
16. **Huberman, E., Sachs, L., Yang, S. K., and Gelboin, H. V.,** Identification of mutagenic metabolites of benzo(a)pyrene in mammalian cells, *Proc. Natl. Acad. Sci. U.S.A.*, 73, 607, 1976.

17. **Levin, W., Wood, A. W., Yagi, H., Jerina, D. M., and Conney, A. H.,** (±)-*trans*-7,8-di-hydroxy-7,8-dihydrobenzo(a)pyrene: a potent skin carcinogen when applied topically to mouse skin, *Proc. Natl. Acad. Sci. U.S.A.*, 73, 3867, 1976.
18. **King, H. W. S., Osborne, M. R., Beland, F. A., Harvey, R. G., and Brookes, P.,** (±)-7α, 8β-dihydroxy-9β, 10β-epoxy-7,8,9,10-tetra-hydrobenzo(a)pyrene is an intermediate in the metabolism and binding to DNA of benzo(a)pyrene, *Proc. Natl. Acad. Sci. U.S.A.*, 73, 2679, 1976.
19. **Hulbert, P. B.,** Carbonium ion as ultimate carcinogen of polycyclic aromatic hydrocarbons, *Nature (London)*, 256, 146, 1975.
20. **Dipple, A., Lawley, P. D., and Brookes, P.,** Theory of tumour initiation by chemical carcinogens: dependence of activity on structure of ultimate carcinogen, *Eur. J. Cancer*, 4, 493, 1968.
21. **Lehr, R. E., Kumar, S., Levin, W., Wood, A. W., Chang, R. L., Conney, A. H., Yagi, H., Sayer, J. M., and Jerina, D. M.,** The bay region theory of polycyclic aromatic hydrocarbon carcinogenesis, in *Polycyclic Hydrocarbons and Carcinogenesis*, Harvey, R. G., Ed., American Chemical Society, Washington, D.C., 1985, 63.
22. **Thakker, D. R., Yagi, H., Akagi, H., Koreeda, M., Lu, A. Y. H., Levin, W., Wood, A. W., Conney, A. H., and Jerina, D. M.,** Metabolism of benzo(a)pyrene. VI. Stereoselective metabolism of benzo(a)pyrene and benzo(a)pyrene and benzo(a)pyrene 7,8-dihydrodiol to diol epoxides. *Chem. Biol. Interact.*, 16, 281, 1977.

22a. **Sayer, J. M., Yagi, H., Silverton, J. V., Friedman, S. L., Whalen, D. L., and Jerina, D. M.,** Conformational effects in the hydrolysis of rigid benzylic epoxides: implications for diol epoxides of polycyclic hydrocarbons, *J. Am. Chem. Soc.*, 104, 1972, 1982.

23. **Beland, F. A. and Harvey, R. G.,** The isomeric 9, 10-oxides of *trans*-7,8-dihydroxy-7,8-dihydrobenzo(a)pyrene, *J. Chem. Soc. Chem. Commun.*, p. 84. 1976.
24. **Yang, S. K., McCourt, D. W., Roller, P. P., and Gelboin, H. V.,** Enzymatic conversion of benzo(a)pyrene leading predominantly to the diol-epoxide r-7, t-8-dihydroxy-t-9,10-oxy-7,8,9,10-tetrahydrobenzo(a)pyrene through a single enantiomer of r-7, T-8-dihydroxy-7,8-dihydrobenzo(a)pyrene, *Proc. Natl. Acad. Sci. U.S.A.*, 73, 2594, 1976.
25. **Kapitulnik, J., Wislocki, P. G., Levin, W., Yagi, H., Thakker, D. R., Akagi, H., Koreeda, M., Jerina, D. M., and Conney, A. H.,** Tumorigenicity studies with diol-epoxides of benzo(a)pyrene which indicate that (±)-*trans*-7β, 8α-dihydroxy-9α, 10α-epoxy-7,8,9,10-tetra-hydrobenzo(a)pyrene is an ultimate carcinogen in newborn mice, *Cancer Res.*, 38, 354, 1978.
26. **Buening, M. K., Wislocki, P. G., Levin, W., Yagi, H., Thakker, D. R., Akagi, H., Koreeda, M., Jerina, D. M., and Conney, A. H.,** Tumorigenicity of the optical enantiomers of the diastereomeric benzo(a)pyrene 7,8-diol-9,10-epoxides in newborn mice: exceptional activity of (+)-7β, 8α, dihydroxy-9α, 10α-epoxy-7,8,9,10-tetrahydrobenzo(a)pyrene, *Proc. Natl. Acad. Sci. U.S.A.*, 75, 5358, 1978.
27. **Yagi, H., Akagi, H., Thakker, D. R., Mah, H. D., Koreeda, M., and Jerina, D. M.,** Absolute stereochemistry of the highly mutagenic 7,8-diol 9,10-epoxides derived from the potent carcinogen trans-7,8-dihydroxy-7,8-dihydrobenzo(a)pyrene, *J. Am. Chem. Soc.*, 99, 2358, 1977.
28. **Yang, S. K., McCourt, D. W., Gelboin, H. V., Miller, J. R., and Roller, P. P.,** Stereochemistry of the hydrolysis products and their acetonides of two stereoisomeric benzo(a)pyrene 7,8-diol 9,10-epoxides, *J. Am. Chem. Soc.*, 99, 5124, 1977.
29. **Whalen, D. L., Ross, A. M., Yagi, H., Karle, J. M., and Jerina, D. M.,** Stereoelectronic factors in the solvolysis of bay region diol epoxides of polycyclic aromatic hydrocarbons, *J. Am. Chem. Soc.*, 100, 218, 1978.
30. **Malaveille, C., Kuroki, T., Sims, P., Grover, P. L., and Bartsch, H.,** Mutagenicity of isomeric diol-epoxides of benzo(a)pyrene and benz(a)anthracene in *S. typhimurium* TA98 and TA100, and in V79 Chinese hamster cells, *Mutat. Res.*, 44, 313, 1977.
31. **Phillips, D. H. and Grover, P. L.,** Biologically-active and chemically-reactive polycyclic hydrocarbon metabolites, in *Monitoring Human Exposure to Carcinogenic and Mutagenic Agents*, Berlin, A., Draper, M., Hemminki, K., and Vainio, H., Eds., International Agency for Research on Cancer, Lyon, 1984, 47.
32. **Dumitz, J. D.,** *X-ray Analysis and the Structure of Organic Molecules*, Connell University Press, Ithaca, 1979.
33. **Horn, A. S. and De Ranter, C. J., Eds.,** *X-ray Crystallography and Drug Action*, Clarendon Press, Oxford, 1984.
34. **Stout, G. H. and Jensen, L. H.,** *X-ray Structure Determination*, Macmillan, New York, 1968.
35. **Ladd, M. F. C. and Palmer, R. A.,** *Structure Determination by X-ray Crystallography*, 2nd ed., Plenum Press, New York, 1985.
36. **Allen, F. H., Bellard, S., Brice, M. D., Cartwright, B. A., Doubleday, A., Higgs, H., Hummelink, T., Hummelink-Peters, B. G., Kennard, O., Motherwell, W. D. S., Rodgers, J. R., and Watson, D. G.,** The Cambridge Crystallographic Data Centre: computer-based search, retrieval, analysis and display of information, *Acta Crystallogr.*, B35, 2331, 1979.

37. **Zacharias, D. E., Glusker, J. P., Fu, P. P., and Harvey, R. G.,** Molecular structures of the dihydrodiols and diol epoxides of carcinogenic polycyclic aromatic hydrocarbons. X-ray crystallographic and NMR analysis, *J. Am. Chem. Soc.*, 101, 4043, 1979.
38. **Neidle, S., Subbiah, A., and Osborne, M. R.,** The molecular structure of (±)-7,8-dihydroxy-7,8-dihydrobenzo(a)pyrene, an early metabolite of benzo(a)pyrene, *Carcinogenesis,* 2, 533, 1981.
39. **Zacharias, D. E., Glusker, J. P., Harvey, R. G., and Fu, P. P.,** Molecular structure of the K-region cis-dihydrodiol of 7,12-dimethylbenz(a)anthracene, *Cancer Res.*, 37, 775, 1977.
40. **Neidle, S., Subbiah, A., Kuroda, R., and Cooper, C. S.,** Molecular Structure of (±)-7,8,9,10-tetrahydroxy-7,8,9,10-tetrahydrobenzo(a)pyrene determined by X-ray crystallography, *Cancer Res.*, 42, 3766, 1982.
41. **Neidle, S., Cooper, C. S., and Ribeiro, O.,** The molecular structure of (±)-8α,9β, 10β, 11α-tetrahydroxy-8,9,10,11-tetrahydrobenz(a)anthracene; an X-ray crystallographic analysis, *Carcinogenesis,* 2, 445, 1981.
42. **Glusker, J. P., Carrell, H. L., Zacharias, D. E., and Harvey, R. G.,** Crystallographic studies of K-region arene oxides: 7,12-dimethylbenz(a)anthracene-5,6-oxide and phenanthrene-9,10-oxide, *Cancer Biochem. Biophys.*, 1, 43, 1974.
43. **Glusker, J. P., Zacharias, D. E., Carrell, H. L., Fu, P. P., and Harvey, R. G.,** Molecular structure of benzo(a)pyrene 4,5-oxide, *Cancer Res.*, 36, 3951, 1976.
44. **Kuroda, R. and Neidle, S.,** Molecular structure of (±)-3-epoxycyclopenta (cd) pyrene; an x-ray crystallographic study, *Carcinogenesis,* 4, 217, 1983.
45. **Neidle, S., Subbiah, A., Cooper, C. S., and Riberio, O.,** Molecular structure of (±)-7α, 8β-dihydroxy-9β, 10β-epoxy-7,8,9,10-tetrahydro-benzo(a)pyrene; an X-ray crystallographic study, *Carcinogenesis,* 1, 249, 1980.
46. **Neidle, S. and Cutbush, S. D.,** X-ray crystallographic analysis of (±)-7β, 8α-dihydroxy-9β, 10β-epoxy-7,8,9,10-tetrahydro-benzo(a)pyrene: molecular structure of a *'syn'* diol epoxide, *Carcinogenesis,* 4, 415, 1983.
47. **Klein, C. L. and Stevens, E. D.,** Molecular structure of *anti*-3,4-dihydroxy-1,2,3,4-tetrahydronaphthalene 1,2-oxide, *Cancer Res.*, 44, 1523, 1984.
48. **Klein, C. L. and Stevens, E. D.,** Structure of *syn*-3,4-dimethoxy-1,2-epoxy-1,2,3,4,-tetrahydronaphthalene, *Acta Crystallogr.*, C40, 315, 1984.
49. **Glusker, J. P., Zacharias, D. E., Whalen, D. L., Friedman, S., and Pohl, T. M.,** Internal hydrogen bond formation in a *syn*-hydroxyepoxide, *Science,* 215, 695, 1982.
50. **Beland, F. A., Melchior, W. B., Jr., Klimkowski, V. J., Scarsdale, J. N., Van Alsenoy, C., and Schafer, L.,** Ab initio study on the molecular structure of the naphthalene metabolite, trans-1,2-dihydroxy-1,2-dihydronaphthalene, *Carcinogenesis,* 5, 1097, 1984.
51. **Allen, F. H.,** The geometry of small rings. V. Geometric variations and hybridisation in three-membered heterocycles C_2X (X=N,O,Si,P,S) and CXY(X,Y=N,O) *Tetrahedron,* 38, 2843, 1982.
52. **Silverman, B. D.,** Determination of the molecular structure of the bay-region epoxides of benzo(a)pyrene by a force-field calculation, *Cancer Biochem. Biophys.*, 6, 131, 1983.
53. **Klopman, G., Ginnberg, H., and Hopfinger, A. J.,** MINDO/3 calculations of the conformation and carcinogenicity of epoxy-metabolites of aromatic hydrocarbons: 7,8-dihydroxy-9,10-oxy-7,8,9,10-tetrahydrobenzo(a)pyrene, *J. Theor. Biol.*, 79, 355, 1979.
54. **Yeh, C. Y., Fu, P. P., Beland, F. A., and Harvey, R. G.,** Application of CNDO/2 theoretical calculations to interpretation of the chemical reactivity and biological activity of the syn and anti diolepoxides of benzo(a)pyrene, *Bioorg. Chem.*, 7, 497, 1978.
55. **Kikuchi, O., Hopfinger, A. J., and Klopman, G.,** Electronic structure and reactivity of four stereoisomers of benzo(a)pyrene 7,8-diol-9,10-epoxide, *Cancer Biochem. Biophys.*, 4, 1, 1979.
56. **Yagi, H., Hernandez, O., and Jerina, D. M.,** Synthesis of (±)-7β, 8α-dihydroxy-9β, 10β-epoxy-7,8,9,10-tetrahydrobenzo(a) pyrene with stereochemistry related to the antileukeoic triptolides, *J. Am. Chem. Soc.*, 97, 6881, 1975.
57. **Whalen, D. L., Montemarano, J. A., Thakker, D. R., Yagi, H., and Jerina, D. M.,** Changes of mechanisms and product distributions in the hydrolysis of benzo(a)pyrene-7,8-diol-9,10-epoxide metabolites induced by changes in pH, *J. Am. Chem. Soc.*, 99, 5522, 1977.
58. **Sayer, J. M., Yagi, H., Silverton, J. V., Friedman, S. L., Whalen, D. L., and Jerina, D. M.,** Conformational effects in the hydrolysis of rigid benzylic epoxides: implications for diol epoxides of polycyclic hydrocarbons, *J. Am. Chem. Soc.*, 104, 1972, 1982.
59. **Sayer, J. M., Whalen, D. L., Friedman, S. L., Paik, A., Yagi, H., Vyas, K. P., and Jerina, D. M.,** Conformational effects in the hydrolyses of benzo-ring diol epoxides that have bay-region diol groups, *J. Am. Chem. Soc.*, 106, 226, 1984.
60. **Murray-Rust, P. and Glusker, J. P.,** Directional hydrogen bonding to sp^2- and sp^3-hybridized oxygen atoms and its relevance to ligand-macromolecule interactions, *J. Am. Chem. Soc.*, 106, 1018, 1984.
61. **Carrell, H. L., Glusker, J. P., Moschel, R., Hudgins, W. R., and Dipple, A.,** Crystal structure of carcinogen: nucleoside adduct, *Cancer Res.*, 41, 2230, 1981.

62. **Stezowski, J. J., Stigler, R.-D., Joos-Guba, G., Kahre, J., Lösch, G. R., Cappell, H. L., Peck, R. M., and Glusker, J. P.,** Stereochemical properties of nucleosides alkylated by activated carcinogens, *Cancer Res.*, 44, 555, 1984.
63. **Lawley, P. D. and Thatcher, C. J.,** Methylation of deoxyribonucleic acid in cultured mammalian cells by N-methyl-N-nitro-N-nitrosoguanidine. The influence of cellular thiol concentrations on the extent of methylation and the 6-oxygen atom of guanine as a site of methylation, *Biochem. J.*, 116, 693, 1976.
64. **Royer, R. E., Lyle, T. A., Moy, G. G., Daub, G. H., and Vander Jagt, D. L.,** Reactivity-selectivity properties of reactions of carcinogenic electrophiles with biomolecules: kinetics and products of the reaction of benzo(a)pyrene-6-methyl cation with nucleosides and deoxynucleosides, *J. Org. Chem.*, 4, 3202, 1979.
65. **Loveless, H.,** Possible relevance of O-6 alkylation of deoxyguanosine to the mutagenicity and carcinogenecity of nitrosamines and nitrosamides, *Nature (London)*, 223, 206, 1969.
66. **Osterman-Golkar, S., Ehrenberg, L., and Wachtmeister, C. A.,** Reaction kinetics and biological action in barley of monofunctional methanesulfonic esters, *Radiat. Bot.*, 10, 303, 1970.
67. **Moschel, R. C., Hudgin, W. R., and Dipple, A.,** Selectivity in nucleoside alkylation and aralkylation in relation to chemical carcinogenesis, *J. Org. Chem.*, 44, 3324, 1979.
68. **Moschel, R. C., Hudgins, W. C., and Dipple, A.,** Alkylation of guanosine by the carcinogen N-nitro-N-benzylurea, *J. Orig. Chem.*, 45, 533, 1980.
69. **Osborne, M. R., Beland, F. A., Harvey, R. G., and Brookes, P.,** The reaction of (±)-7α, 8β-dihyhoxy 9β,10β-epoxy 7,8,9,10-tetrahydrobenzo(a)pyrene with DNA, *Int. J. Cancer*, 18, 362, 1976.
70. **Weinstein, J. B., Jeffrey, A. M., Jennette, K. W., Blobstein, S. H., Harvey, R. G., Kasai, H., and Nakanishi, K.,** Benzo(a)pyrene diol epoxides as intermediates in nucleic acid binding *in vitro* and *in vivo*, *Science*, 193, 592, 1976.
71. **Jeffrey, A. M., Jennette, K. W., Blobstein, S. H., Weinstein, J. B., Beland, F. A., Harvey, R. G., Kasai, H., Miura, I., and Nakanishi, K.,** Benzo(a)pyrene-nucleic acid derivative found *in vivo:* structure of benzo(a)pyrene tetrahydrodiol epoxide-guanosine adduct, *J. Am. Chem. Soc.*, 98, 5714, 1976.
72. **Koreeda, M., Moore, P. D., Yagi, H., Yeh, H. C., and Jerina, D. M.,** Alkylation of polyguanylic acid at the 2-amino group and phosphate by the potent mutagen (±)-7,8-dichydroxy 9,10-epoxy 7,8,9,10-mutagen (±)-7,8,9,10-tetrahydrobenzo(a)pyrene, *J. Am. Chem. Soc.*, 98, 6720, 1976.
73. **Straub, K. M., Meehan, T., Burlingame, A. L., and Calvin, M.,** Identification of the major adducts formed by reaction of benzo(a)pyrene diolepoxide with DNA *in vitro*, *Proc. Natl. Acad. Sci. U.S.A.*, 74, 5285, 1977.
74. **Meehan, T., Straub, K., and Calvin, M.,** Benzo(a)pyrene diol epoxide covalently binds to deoxyguanosine and deoxyadenosine in DNA, *Nature (London)*, 269, 725, 1977.
75. **Osborne, M. R., Jacobs, S., Harvey, R. G., and Brookes, P.,** Minor products from the reaction of (+) and (−) benzo(a)pyrene-anti-diolepoxide with DNA, *Carcinogenesis*, 2, 533, 1981.
76. **King, H. W. S., Osborne, M. R., and Brookes, P.,** The *in vivo* and *in vitro* reaction at the N7-position of guanine of the ultimate carcinogen derived from benzo(a)pyrene, *Chem. Biol. Interact.*, 24, 345, 1979.
77. **Osborne, M. R., Harvey, R. G., and Brookes, P.,** The reaction of *trans*-7,8-dihydroxy *anti*-9,10-epoxy 7,8,9,10-tetrahydrobenzo(a)pyrene with DNA involves attack at the N7-position of guanine moieties, *Chem. Biol. Interact.*, 20, 123, 1978.
78. **Meehan, T. and Straub, K.,** Double-stranded DNA stereoselectively binds benzo(a)pyrene diol epoxides, *Nature (London)*, 277, 410, 1979.
79. **Wood, A. W., Chang, R. L., Levin, W., Yagi, H., Thakker, D. R., Jerina, D. M., and Conney, A. H.,** Differences in mutagenicity of the optical enantiomers of the diastereoisomeric benzo(a)pyrene 7,8-diol-9,10-epoxides, *Biochem. Biophys. Res. Commun.*, 17, 1389, 1977.
80. **Lefkowitz, S. M., Brenner, H. C., Astorian, D. G., and Clarke, R. H.,** Optically detected magnetic resonance study of benzo(*a*)pyrene-7,8-dihydrodiol 9,10-oxide covalently bound to DNA, *FEBS Lett.*, 105, 77, 1979.
81. **Prusik, T., Geacintov, N. E., Tobiasz, C., Ivanovic, V., and Weinstein, J. B.,** Fluorescence study of the physico-chemical properties of a benzo(a)pyrene 7,8-dihydrodiol 9,10-oxide derivative bound covalently to DNA, *Photochem. Photobiol.*, 29, 223, 1979.
82. **Prusik, T. and Geacintov, N. E.,** Fluorescence properties of a benzo(a)pyrene 7,8 dihydrodiol 9,10-oxide-DNA adduct. Conformation and effects of intermolecular DNA interactions, *Biochem. Biophys. Res. Commun.*, 88, 782, 1979.
83. **Drinkwater, N. R., Miller, J. A., Miller, E. C., and Yang, N.-C.,** Covalent intercalative binding to DNA in relation to the mutagenicity of hydrogen epoxides and N-acetoxy-2-acetylaminofluorene, *Cancer Res.*, 38, 3247, 1978.
84. **Ibanez, V., Geacintov, N. E., Gagliano, A. G., Brandimarte, S., and Harvey, R. G.,** Physical binding of tetrols derived from 7,8-dihydroxy-9,10-epoxybenzo(a)pyrene to DNA, *J. Am. Chem Soc.*, 102, 5661, 1980.

85. **Geacintov, N. E., Gagliano, A. G., Ibanez, V., and Harvey, R. G.,** Spectroscopic characterizations and comparisons of the structures of the covalent adducts derived from the reactions of 7,8-dihydroxy-7,8,9,10-tetrohydrobenzo(a)pyrene 9,10-oxide, and the 9,10-epoxides of 7,8,9,10-benzo(a)pyrene and 9,10,11,12-tetrahydrobenzo(a)pyrene with DNA, *Carcinogenesis,* 3, 247, 1982.
86. **Geacintov, N. E.,** Mechanism of interaction of polycyclic aromatic diol epoxides with DNA and structures of the adducts, *ACS Symp. Ser.,* Vol. 283, Comstock, M. J. Ed., American Chemical Society, Washington, D.C., 1985, 107.
87. **Geacintov, N. E., Gagliano, A., Ivanovic, V., and Weinstein, I. B.,** Electric dichroism study on the orientation of benzo(a)pyrene-7,8-dihydrodiol-9,10-oxide covalently bound to DNA, *Biochemistry,* 17, 5256, 1978.
88. **Undeman, O., Lycksell, P. O., Graslund, A., Astlind, T., Ehrenberg, A., Jernstrom, B., Tjerneld, F., and Norden, B.,** Covalent complexes of DNA and two stereoisomers of benzo(a)pyrene 7,8-dihydrodiol-9,10-epoxides studied by fluorescence and linear dichroism, *Cancer Res.,* 43, 1851, 1983.
89. **Meehan, T., Gamper, H., and Becker, J. F.,** Characterization of reversible, physical binding of benzo(a)pyrene derivatives to DNA, *J. Biol. Chem.,* 257, 10479, 1982.
90. **Geacintov, N. E., Yoshida, H., Ibanez, V., and Harvey, R. G.,** Non-covalent intercalative binding of 7,8-dihydroxy 9,10-epoxybenzo(a)pyrene to DNA, *Biochem. Biophys. Res. Commun.,* 100, 1569, 1981.
91. **Geacintov, N. E., Yoshida, H., Ibanez, V., Jacobs, S. A., and Harvey, R. G.,** Conformations of adducts and kinetics of binding to DNA of the optically pure enantiomers of anti-benzo(a)pyrene diol epoxide, *Biochem. Biophys. Res. Commun.,* 122, 33, 1984.
92. **Geacintov, N. E., Ibanez, V., Gagliano, A. G., Jacobs. S. A., and Harvey, R. G.,** Stereoselective covalent binding of anti-benzo(a)pyrene diol epoxide to DNA, conformation of enantiomer adducts, *J. Biomolec. Struct. Dyn.,* 1, 1473, 1984.
93. **Gamper, H. B., Straub, K., Calvin, M., and Bartholomew, J. C.,** DNA alkylation and unwinding induced by benzo(a)pyrene diol epoxide: modulation by ionic strength and superhelicity, *Proc. Natl. Acad. Sci. U.S.A.,* 77, 2000, 1978.
94. **Kakefuda, T. and Yamamoto, H.-A.,** Modification of DNA by the benzo(a)pyrene metabolite diol-epoxide r-7,t-8-dihydroxy-t-9,10-oxy-7,8,9,10-tetrahydrobenzo(a)pyrene, *Proc. Natl. Acad. Sci. U.S.A.,* 75, 415, 1978.
95. **Gamper, H. B., S.-C., Tung, A., Straub, K., Bartholomew, J., and Calvin, M.,** DNA strand scission by benzo(a)pyrene diol epoxides, *Science,* 197, 671, 1977.
96. **Agarwal, K. L., Hrinyo, T. P., and Yang, N.-C. C.,** Interaction of (±)-7r,8t-dihydroxy-9t,10t-epoxy-7,8,9,10-tetrahydrobenzo(a)pyrene with relaxed circular pBR322 DNA, *Biochem. Biophys. Res. Commun.,* 114, 14, 1983.
97. **Frenkel, K., Grunberger, D., Boublik, M., and Weinstein, I. B.,** Conformation of dinucleoside monophosphates modified with benzo(a)pyrene-7,8-dihydrodiol 9,10-oxide as measured by circular dichoism, *Biochemistry,* 17, 1278, 1978.
98. **Pulkrabek, P., Leffler, S., Weinstein, I. B., and Grunberger, D.,** Conformation of DNA modified with a dihydrodiol epoxide derivatives of benzo(a)pyrene, *Biochemistry,* 16, 3127, 1977.
99. **Hogan, M. E., Dattaguputa, N., and Whitlock, J. P.,** Carcinogen-induced alteration of DNA structure, *J. Biol. Chem.,* 256, 4504, 1981.
100. **Hingerty, B. and Broyde, S.,** A conformational analysis of the (+) anti BPDE adduct to the guanine amino group of dCpdG, *J. Biomolec. Struct. Dyn.,* 1, 905, 1983.
101. **Glusker, J. P.,** X-ray analyses of polycyclic hydrocarbon metabolite structures, *ACS Symp. Ser.,* Vol. 283, Constock, M. J. Ed., American Chemical Society, Washington, D.C., 1985, 125.
102. **Blobstein, S. H., Weinstein, I. B., Grunberger, D., Weisgras, J. W., and Harvey, R. G.,** Products obtained after in vitro reaction of 7,12-dimethyl-benzo(a)anthracene 5,6-oxide with nucleic acids, *Biochemistry,* 14, 3451, 1975.
103. **Jeffrey, A. M., Blobstein, S. H., Weinstein, I. B., Beland, F. A., Harvey, R. G., Kasai, H., and Nakanishi, K.,** Structure of 7,12-dimethyl-benz(a)anthracene-guanine adducts, *Proc. Natl. Acad. Sci. U.S.A.,* 73, 2311, 1976.
104. **Taylor, E. R., Miller, K. J., and Bleyer, A. J.,** Interactions of molecules with nucleic acids. X. Covalent intercalative binding of the carcinogenic BPDE I(+) to kinked DNA, *J. Biomolec. Struct. Dyn.,* 1, 883, 1983.
105. **Islam, S. A., Neidle, S., Gandecha, B. M., and Brown, J. R.,** Experimental and computer graphics simulation analyses of the DNA interaction of 1,8-bis-(2-diethylaminoethylamino)-anthracene-9,10-dione, a compound modelled on doxorubicin, *Biochem. Pharmacol.,* 32, 2801, 1983.
106. **Gabbey, E. J., DeStefano, R., and Baxter, C. S.,** Topography of nucleic acid helices in solutions. XXVIII. Evidence for a dynamic structure of DNA in solution, *Biochem. Biophys. Res. Commun.,* 51, 1083, 1973.

107. **Miller, K. J., Taylor, E. R., and Dommen, J.,** A mechanism for stereoselectivity and binding of benzo(a)pyrene diol epoxides to DNA in *ACS Symp. Ser.*, Vol. 283, Comstock, M. J., Ed., American Chemical Society, Washington, D.C., 1985, 239.
108. **Beland, F. A.,** Computer-generated graphic models of the N^2-substituted deoxyguanosine adducts of 2-acetylaminofluorene and benzo(a)pyrene and the O^6-substituted deoxyguanosine adduct of 1-naphthylamino in the DNA double helix, *Chem. Biol. Interact.*, 22, 329, 1978.
109. **Arnott, S., Campbell-Smith, R. J., and Chandrasekaran, R.,** in *Handbook of Biochemistry and Molecular Biology*, Vol. 2, 3rd ed., Fasman, G. D., Ed., CRC Press, Boca Raton, Fla., 1976, 411.
110. **Aggarwal, A. K., Islam, S. A., and Neidle, S.,** Computer modelling studies of the covalent interactions between DNA and the enantiomers of anti-7,8-diol,9,10-epoxy-benzo(a)pyrene, *J. Biomolec. Struct. Dyn.*, 873, 1983.
111. **Taylor, E. and Miller, K. J.,** Interactions of molecules with nucleic acids. IX. Generalization of techniques to generate nucleic acid structures with applications to intercalation sites and kinked structures, *Biopolymers,* 23, 2853, 1984.
112. **Lerman, L. S.,** Structural considerations in the interaction of DNA and acridines, *J. Mol. Biol.*, 3, 18, 1961.
113. **Shieh, H. S., Berman, H. M., Dabrow, M., and Neidle, S.,** The structure of drug-deoxydinucleoside phosphate complex: generalized conformational behaviour of intercalation complexes with RNA and DNA fragments, *Nucleic Acids Res.*, 8, 85, 1980.
114. **Berman, H. M., Stallings, W., Carrell, H. L., Glusker, J. P., Neidle, S., Taylor, G., and Achari, A.,** Molecular and crystal structure of an intercalation complex: proflavine-cytidylyl-(3′,5′)-guanosine, *Biopolymers,* 18, 2405, 1979.
115. **Tsai, C.-C., Jain, S. C., and Sobell, H. M.,** Visualization of drug-nucleic acid interactions at atomic resolution. I. Structure of an ethidium/dinucleoside monophosphate crystalline complex, ethidium: 5-iodouridyl(3′-5′)adenosine, *J. Mol. Biol.*, 114, 301, 1977.
116. **Jain, S. C., Tsai, C.-C., and Sobell, H. M.,** Visualization of drug-nucleic acid interactions at atomic resolution. II. Structure of an ethidium/dinucleoside monophosphate crystalline complex ethidium: 5-iodoeytidyl(3′-5′)guanosine, *J. Mol. Biol.*, 114, 317, 1977.
117. **Reddy, B. S., Seshadri, T. P., Sakore, T. D., and Sobell, H. M.,** Visualization of drug-nucleic acid interactions at atomic resolution. V. Structures of two aninoacridine-dinucleoside monophosphate crystalline complexes; proflavin-5-iodocytidyl (3′-5′) guanosine and acridine orange-5-idocytidylyl(3′-5′)-guanosine, *J. Mol. Biol.*, 135, 787, 1979.
118. **Subbiah, A., Islam, S. A., and Neidle, S.,** Molecular modelling studies on the non-covalent intercalative interactions between DNA and the enantiomers of anti-benzo(a)pyrene 7,8-diol-9,10-epoxide, *Carcinogenesis,* 4, 211, 1983.
119. **Quigley, G. J., Wang, A. H.-J., Ughetto, G., van der Marel, G., van Boom, J. H., and Rich, A.,** Molecular structure of an anticancer drug-DNA complex: daunomycin plus d(CpGpTpApCpG), *Proc. Natl. Acad. Sci. U.S.A.*, 77, 7204, 1980.

INDEX

A

B

C

D

E

F

G

H

I

J

K

L

M

N

O

P

Q

R

S

T

U

V

W

X

Y

Z